Looking for a Few Good Males

ANIMALS, HISTORY, CULTURE

Harriet Ritvo, Series Editor

Looking for a Few Good Males

FEMALE CHOICE IN EVOLUTIONARY BIOLOGY

ERIKA LORRAINE MILAM

The Johns Hopkins University Press

Baltimore

© 2010 The Johns Hopkins University Press
All rights reserved. Published 2010
Printed in the United States of America on acid-free paper
2 4 6 8 9 7 5 3 1

The Johns Hopkins University Press
2715 North Charles Street
Baltimore, Maryland 21218-4363
www.press.jhu.edu

Library of Congress Cataloging-in-Publication Data

Milam, Erika Lorraine, 1974–
Looking for a few good males : female choice in evolutionary biology /
Erika Lorraine Milam.
p. cm. — (Animals, History, Culture)
Includes bibliographical references and index.
ISBN-13: 978-0-8018-9419-0 (hardcover : alk. paper)
ISBN-10: 0-8018-9419-0 (hardcover : alk. paper)
1. Courtship in animals. 2. Evolution (Biology). 3. Courtship in animals—
Research—History—20th century. I. Title.
QL761.M48 2010
591.56'2—dc22 2009023833

A catalog record for this book is available from the British Library.

*Special discounts are available for bulk purchases of this book. For more information,
please contact Special Sales at 410-516-6936 or specialsales@press.jhu.edu.*

The Johns Hopkins University Press uses environmentally friendly book materials,
including recycled text paper that is composed of at least 30 percent post-consumer
waste, whenever possible. All of our book papers are acid-free, and our jackets
and covers are printed on paper with recycled content.

For my family, both old and new, near and far

CONTENTS

Looking for a Few Good Males

Introduction

How does a hen decide whether she is in the mood for sex? How does a hen decide which is the best rooster? For a female chicken to "choose" a rooster, in the way we think of humans choosing partners, she would have to recognize differences between males, compare them based on that variation, and decide with which male(s) she should mate and invest her reproductive future. Such cognitive comparisons imply a level of rational thought and aesthetic appreciation that many biologists in the twentieth century were hesitant to ascribe to nonhuman animals. For these biologists, rational choice was reserved for humans, and to interpret the behavior of animals in choice-based terms could be described only as anthropomorphic. Yet, despite this criticism, other biologists have been fascinated by the idea that the mating behavior of animals could serve as an experimental entrée to the more complicated process of mate choice in humans. In both cases, the history of female choice in evolutionary biology reflects biologists' careful negotiation of the implied aesthetics and rationality of choice-based behavior in humans and other animals.

The touchstone for research on female mate choice continues to be Charles Darwin's theory of sexual selection. In 1871, Darwin published *Descent of Man and Selection in Relation to Sex*, in which he elaborated his "second" theory.[1] Sexual selection consisted of two mechanisms that together explained why males and females differed in appearance and behavior: one was female choice, the other was male-male competition. In female choice, females compared the mating displays of reproductively mature males and chose the most appealing male with which to mate. In male-male competition, males fought to determine which male would have access to the reproductively available female(s) in the area. Both mechanisms, Darwin argued, would result in exaggerated male traits. Over evolutionary time, female choice would lead to aesthetically pleasing male traits (long tails and brightly colored plumage), and male-male competition to armor or weapons (bony plates or antlers).

Darwin contrasted sexual selection with his more widely known theory of natural selection; whereas natural selection depended on who survived, sexual selection depended on who reproduced. Natural selection would lead to utilitarian adaptations. Sexual selection through female choice provided Darwin with an explanation for a variety of aesthetic phenomena he could not explain through survival alone: the presence of beauty in animals, differences between males and females of the same species, and racial differences within a species. Thus, for late nineteenth-century biologists, female choice presupposed both a sense of aesthetic appreciation and an ability to choose rationally based on this aesthetic sensibility—mental attributes they were hesitant to ascribe to animal minds. In the animal-human cognitive divide, humans were capable of rational aesthetic choices, and animals were not.

According to most historical accounts of sexual selection theories, it was not until the 1970s that organismal field biologists self-consciously returned to a Darwinian model of female choice as a mechanism for evolutionary change, one that emphasized the continuity of rationality and choice across the animal-human boundary.[2] Historians and biologists alike typically identify female choice as the cause of the eclipse of interest in sexual selection, and they point to Robert Trivers's monumental paper "Parental Investment and Sexual Selection," published in 1972, as inaugurating new interest in the theory by providing a biological justification for female choice in nature.[3] Certainly, his paper had been cited more than three thousand times by 2000 (and over fifteen hundred more times by 2008), a citation record that speaks to the remarkable interest in sexual selection as a topic of research in recent decades.

The history of mate choice is distinct from the history of sexual selection as an explanation of beauty in the animal kingdom. Although biologists' interest in sexual selection did wane for much of the twentieth century, interest in female choice remained strong. In the early decades of the twentieth century, biologists reframed female choice in the mechanistic language of "stimulation" rather than the aesthetic, cognitive language of "choice." Apparent mate choice was really due to the differential stimulation provided by the courtship displays of potential suitors. When choice-based behaviors in animals were recast within a stimulatory framework, they survived; when "choice" implied cognitive comparison, as in the aesthetic evaluation of beauty, they did not. Inasmuch as female choice was an issue for biologists between Darwin and Trivers, biologists' concerns centered on anthropomorphic "choice" rather than "female" agency.

If history is written by the victors, then the characters in the intervening decades were losers merely in the sense that the historical packaging of sexual selection and female choice has either painted their research as overlooked and forgotten by their contemporaries (when indeed it was not) or entirely excluded them from the history. In one sense, this book is a recovery of those forgotten elements of the history of sexual selection. In another sense, it is a history of how changing conceptions of animals' minds affected not only the scientific reception of sexual selection and female choice as theories but also the ways in which the histories of these theories have been negotiated.

In the early decades of the twentieth century, both biologists and psychologists considered choice-based behavior to be far more likely in humans than in animals, as animals seemed to lack the cognitive ability to discern the minute aesthetic differences distinguishing one potential mate from another. Whereas Darwin hoped to access the inner lives of animals through an analysis of their behavior and expressions, his belief in the continuity of animal and human mind became increasingly problematic with the blossoming of Mendelian genetics as a field of inquiry and with a renewed emphasis on controlling animals in a laboratory setting.[4] Although biologists rejected notions of choice in animal courtship, people with an interest in applying evolutionary theory to human affairs continued to embrace sexual selection by female choice. Female choice in marriage partners appeared to be a powerful evolutionary mechanism with which to improve the evolutionary future of humanity. Human females' choice in reproductive partners determined the value of their offspring, in terms of both the quantity and quality of children.[5] Within an evolutionary framework, *minds* and aesthetic sensibilities continued to be discontinuous elements demarcating animal from human, even as scientists simultaneously described *bodies* in terms of continuous physical evolution.

When biologists and mathematicians sought common ground between Mendelian genetics and evolutionary theory, they created an evolutionary model in which the gradual accumulation of minute genetic changes could drive large morphological changes over time.[6] The mathematical models of evolution developed in the 1920s and 1930s reflect this interest in the directional evolution of a single population.[7] Within the new field of what became known as mathematical population genetics, biologists defined evolution as genetic change in a single population over time. Although any given population might improve or degenerate, depending on the circumstances, evolution took place only when the genetic composition of the population was altered; the greater the degree of directional change, the greater the extent of evolution. One mechanism that could drive the evolution of a population was natural selection. Natural selection, to the population geneticists, depended on the total number and quality of offspring an individual produced. This version of natural selection differed sharply from Darwin's vision of natural selection as differential survival and of sexual selection as differential reproduction. By incorporating mate choice into natural selection, the rhetoric of the selecting sex remained important to evolutionary theory during the 1930s, as the primary intent of that theory was to explain humanity's evolutionary past.[8]

A theoretical shift away from a directional model of selection did not occur in evolutionary experiments until later in the 1930s and 1940s, with the synthesis of the new evolutionary theory and naturalists' research on biography, taxonomy, and questions of species identity and distribution.[9] Rather than framing evolution as change within a single population, biologists who had been trained as systematists, paleontologists, and psychologists began to investigate evolution as a process of speciation—the splitting of a single interbreeding population into two reproductively isolated populations. In other words, biologists became less interested in the direction and rate of change of a species along the evolutionary limbs of a tree and, instead, began to concentrate on what happened when one branch split into two. These biologists used mating behavior as a diagnostic tool in determining species identity. If females consistently chose males of their own species with which to mate, then mating behavior could be an important component of maintaining species identity—especially when similar species were found in the same geographic location. Similarly, if some females in a population refused to mate with one kind of male, and other females refused to mate with a second kind of male, then female choice could drive the creation of two reproductively isolated populations (where before there had been only one). In this context, evolutionary biologists no longer considered female mate choice as synonymous with the evolution of a single species, but reframed it as a mechanism that could potentially drive speciation. As the naturalists who forged new programs for evolutionary research described new methods for quantifying animal mating behavior, they deemphasized qualitative descriptions of individual behavior and began to focus on statistical analyses of populations. By downplaying the importance of individual behavior, they dismissed the mental component of mating behavior that had been so integral to behavioral studies before the Second World War.

After the war, biologists became increasingly fascinated by the conjunction of animal behavior and evolutionary theory. Attempts to synthesize these approaches differed in the United Kingdom and the United States in the kinds of questions they asked and the research methodologies they pursued.

Ethology, or the science of studying animal behavior, originated in continental Europe with the work of Konrad Lorenz and Nikolaas Tinbergen, and spread to England with Tinbergen's move to Oxford in 1949.[10] Ethologists suggested that the tendency of females to mate with some males but not others was simply a result of differential male stimulation. Although a female might seem to choose a mate, she was simply responding to the first male who stimulated her sufficiently to overcome her natural tendency not to react. In the 1970s, this model came under fire as couching females as sexually passive and males as active, but such criticisms ignored the point that for ethologists, *all* animals were reacting to stimulation from their environment, all were in some sense passive. Tinbergen and other ethologically trained zoologists were very clear that the most important function of animal courtship was to ensure proper species pairing during the mating season and to stimulate the mating pair into simultaneous sexual interest.

On the other side of the Atlantic, American population geneticists were studying the genetic basis of behavior in fruit flies. Fruit flies seemed so neurologically simple that geneticists discussed the consequences of mate choice in flies without worrying that anyone would think they were implying a cognitive process resembling true human choice. This freedom from anthropomorphic concerns allowed population geneticists to investigate behavior as a possible evolutionary mechanism driving morphological change and rates of speciation. In this research community, female choice and male display were seen as functioning not only to keep species from interbreeding but also to maintain the genetic diversity of a population. When some population geneticists discovered that female flies preferred to mate with whichever male type was found least frequently in a population, they called the phenomenon the "rare-male effect." This ability of female flies to choose different kinds of mates in different situations implied that female flies could, in fact, discern subtle variations among individuals and choose a mate based on those variations. These geneticists argued that through this preference to mate with rare males, female choice acted to maintain the genetic diversity of a population. In this way, the evolutionary mechanism of female choice could act to counter the tendency of natural selection to reduce the genetic diversity of a population.

The success of ethologists and population geneticists in extrapolating their experiments from animals to people was buoyed by the increasing prestige of organismal biology in the 1960s and 1970s. Organismal biology represented a realignment of professional affiliations within the biological sciences. Whereas before the Second World War, zoologists were primarily defined by the identity of their research organism, after the war, biologists became increasingly associated with their methodological approaches—whether ecological, organismal, cellular, or molecular. Although the study of animal behavior became increasingly popular in the 1960s, biologists interested in studying animals under natural conditions felt, in many ways, overshadowed by the incredible advances taking place in molecular biology. The new language of game theory allowed organismal biologists to use the language of choice in animals as shorthand for evolutionarily selected behavioral programs that did not presuppose actual deliberation. At the same time, organismal biologists on both sides of the Atlantic defended their professional ground by claiming intellectual jurisdiction over the scientific study of human social and sexual behavior.

When historically minded biologists began to actively participate in writing the history of their own disciplines, their success in defining sexual selection as an organismal field science corresponded to an erasure of the contributions of theoretical and laboratory-based biologists to the apparent rebirth of scientific interest in female choice in animals and humans. In other words, organismal biologists in the 1970s advanced an eclipse narrative of sexual selection and female choice because that history illustrated the ways in which their current research was innovative. For classical evolutionary biologists such as Ernst Mayr, sexual selection had been eclipsed primarily as a result of the mathematical population geneticists (Fisher, Haldane, and Wright), who redefined natural selection in the 1920s as merely a process of changing gene frequencies. Under such an erroneous definition, to Mayr's mind, individual fitness was measured by the number of genes an individual passed along to the next generation. Any desire to understand why or how such change occurred was obliterated and any mechanism that increased genetic contribution to the next generation was labeled "natural selection," including any mating advantages. Sexual selection was therefore subsumed by natural selection until its "rebirth" in the 1970s. Quantitative theoretical biologists, who had developed mathematical models of female choice as a mechanism of evolutionary change in the 1960s, argued that their research was crucial to ending the eclipse of sexual selection a full decade earlier. Population geneticists, who had begun their research on fruit fly mating in the 1950s and 1960s, similarly argued that the eclipse of interest in female choice had ended earlier than Mayr implied, but in the mid-1950s! These historical narratives reflect the contemporary disciplinary struggles for prestige and funding, and standard histories of female choice and sexual selection stabilized only when the disciplinary identities of biologists interested in animal behavior were also settled. Although each of these research communities interested in females' mate-choice behavior—zoology, mathematical theoretical biology, and experimental population genetics—contributed to the explosion of organismal biologists' interest in female choice and sexual selection in the late 1960s and 1970s, by the 1980s, only the work of nonlaboratory biologists was included in the standard history of sexual selection.

Looking for a Few Good Males explores how the changing landscape of scientific research on animal behavior and evolution affected the ways in which scientists sought to use their knowledge of courtship in animals as the ground for understanding the biological basis of human courtship. Each chapter addresses a separate question that a community of researchers, bounded in space and time, asked themselves when deliberating how to best use studies of nonhuman animal behavior to elucidate human social interactions. Within an evolutionary framework, is human courtship behavior biologically continuous with or discontinuous from the courtship behavior of animals? Which animals possess a cognitive and aesthetic sensibility sufficient to choose a mate? How does the evolutionary process work in nature, and how might courtship behavior change or contribute to evolution as a process? What is the function of an animal's courtship behavior in the context of its evolutionary and ecological history? What is the relationship between natural selection and sexual selection? Who has the scientific authority to pronounce on the relationship between the courtship behavior of animals and humans? These questions were not answered easily, nor were they independent; echoes of each will resurface throughout the book in discussions of female choice and sexual selection. Biologists used their answers to these questions to dispute the boundaries between instinct and learning, between biology and culture, and ultimately between animal and human.

Despite the long history of using animals as models of human behavior, the idea that we could learn about ourselves from studying fruit flies or guppies was far from intuitively obvious and required intellectual work. Biologists used at least two theoretical frameworks, opposite sides of a single coin, to collapse the animal-human boundary—anthropomorphism and zoomorphism. Anthropomorphism, by imbuing animal actions with human emotions or rationality, tended to make animals more like humans. Zoomorphism, by interpreting human behavior as a complex form of animal behavior, or by looking for similarities between the evolutionary functions of human and animal behavior, made humans more like animals.[11] As techniques for conceptualizing human and animal behavior, anthropomorphism and zoomorphism have been around for centuries, and remain deeply important to our modern understanding of the history of female choice in evolutionary and behavioral biology.

By raising issues about the conscious mind and rationality at the animal-human boundary, choice-based mating behavior proved particularly contentious.[12] At one extreme, female choice as an evolutionary mechanism in animals could be anthropomorphic—a female animal "rationally" "compares" mating displays and "chooses" the "prettiest" male with whom to mate. At the other extreme, female choice as an evolutionary mechanism in humans could be zoomorphic—a woman "instinctively" mates with the first man whose courtship "rituals" stimulate her above a minimum excitatory "threshold." Both formulations, conscious choice behavior in animals and the biological basis of behavior in humans, caused long-lasting conceptual disagreements between and within communities of biologists attempting to understand the relationship of behavior and evolution in animals and humans.

The importance of human social and sexual behavior as an object of study within the evolutionary and behavioral sciences fluctuated throughout the twentieth century. In the early decades, biologists concentrated on humans as the direct subject of the effects of female choice as a mechanism for evolutionary change. In the 1940s, the evolution of human behavior largely disappeared as a motivating reason for studying the behavior of animals. Evolutionary biologists became more interested in understanding how evolution worked in natural animal populations at the time, instead of seeking to uncover the past history of behavioral evolution in the animal kingdom, leading inevitably to humans. Contemporaneously, biologists interested in animal behavior shied away from anthropomorphic notions of choice as they strove to increase their professional standing, self-consciously distancing themselves from discussions of human comparisons in their publications for a scientific audience. By the 1960s, however, organismal biologists sought to reclaim disciplinary jurisdiction over the scientific study of human behavior through evolutionary theory. Not surprisingly, their sociobiological theories met with considerable resistance from people who thought that questions of human sexual behavior would be addressed more appropriately by the social sciences. Sociobiology nonetheless became the dominant paradigm within which histories of female choice and sexual selection were written in the 1980s. This return to a human context among field biologists interested in the evolution of social and sexual behavior meant that, from their perspective, no one had been working on these questions in earlier decades, and they did not incorporate laboratory-based research on female choice from the 1940s through the 1960s into their histories of sexual selection. The questions of *whether* to draw a line demarcating a behavioral boundary between animals and humans, and *how* to draw such a line, remain open today.

Do female chickens choose their mates? It depends, as always, on whom you ask, when you ask them, and what you mean by "choice."

CHAPTER 1

Beauty and the Beast

Darwin, Wallace, and the Animal-Human Boundary

What lady gorilla about to become a mother . . . could con-
template without shuddering the probability that, instead of
presenting the gentleman gorilla of her affections with a pledge of
their love that promised to have a hide and a bellow that would
rival those of a buffalo . . . she would produce a wrinkled pink-
bodied weakling, looking like a monkey—one of the smallest and
feeblest of our race—that had been flayed alive, and which, even
after achieving maturity, could only live by covering itself with an
artificial skin, and by making machines with which to get its food
and defend itself against its natural enemies!
—Richard Grant White, *The Fall of Man*, 1871

In describing the "sporting-places" of bower birds in Australia, naturalist John Gould
held forth at considerable length. The bower itself he described as "decorated with
the most gaily colored articles that can be collected, such as the blue tail feathers of
the Rose-hill and Pennantian Parrakeets, bleached bones, the shells of snails, &c."
Gould further suggested that the bowers were "merely sporting-places in which
the sexes meet, and the males display their finery, and exhibit many remarkable
actions."[1] With the birds' elaborate courtship displays and tendencies for interior
decorating, the evolution of bower bird behavior presented a problem for Charles
Darwin's theory of natural selection, because in no way did these behaviors seem to
help the bird fit into and survive in his environment. They appeared, rather, to be
colorful elaborations of an inner aesthetic sensibility.

When Darwin articulated his theory of sexual selection through female choice
and male-male competition, it was ostensibly a theory to explain beauty in the animal kingdom. However, the reception of female choice as an evolutionary mechanism was complicated by both human social politics and concerns over assuming

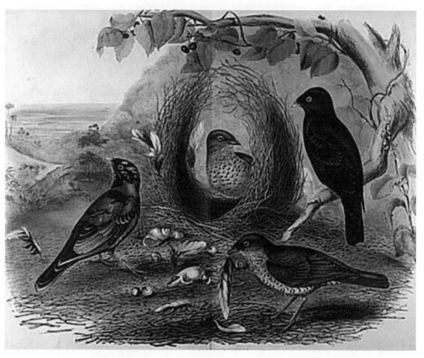

When John Gould arrived in Australia, he was amazed by the playgrounds of the bower bird. He described the bird's antics as follows: "At times the male will chase the female all over the aviary, then go to the bower, pick up a gay feather or a large leaf, utter a curious kind of note, set all his feathers erect, run round the bower . . . he continues opening first one wing and then the other, uttering a low whistling note . . . until at last the female goes gently towards him." John Gould, *Handbook to the Birds of Australia*, vol. 1 (London: Published by the Author, 1865), 444. From Gould, *Birds of Australia*, vol. 4, plate 10, "Satin Bower-Bird."

mentality in animals. Most Victorian scientists deemed nonhuman animals incapable of true choice and dismissed female choice as a promising means of evolutionary change in animals. In humans, however, female choice became a viable, albeit controversial, method of regulating the evolutionary future of society.

Darwin's theory of sexual selection built on his assumptions about normative relations between men and women and the place of Victorian England in the pantheon of great civilizations.[2] Sexual selection provided Darwin with a single answer to two related questions: How was it that males and females of the same species differed significantly in their appearance and behavior? And, how had different races arisen within a species? These questions were related in the sense that both dealt with stable variations within a single species and, for Darwin, were fundamentally questions about the biological origins of the late nineteenth-century society in which he lived. Darwin argued that the psychological continuity of all animal life, including humans, demonstrated that women were intellectually less evolved than men and that all human races were members of the same species.

Darwin's critics and supporters immediately perceived the human context of sexual selection, especially his friend and correspondent Alfred Russel Wallace, whom he recognized as co-discoverer of the theory of natural selection. For both Darwin and Wallace, the issue of female choice in animals called attention to their beliefs about the biological nature of differences between animals and humans. Because Darwin believed in the physical, intellectual, moral, and behavioral continuity of humans with other animals, he applied natural selection and sexual selection with equal weight to both humans and animals. Wallace, however preferred to draw a clear line, with all humans on one side and all animals on the other.[3] He argued that the large gap between animal and human intelligence could have arisen only through the metaphysical intervention of a higher power. Wallace's argument was consistent with his belief that the intellectual and psychological machinery needed for female choice to operate could be found only among humans. By the end of the nineteenth century, many naturalists and psychologists remained unconvinced by either man's arguments.

In describing the reception of Darwin's theory of sexual selection in the late nineteenth century, this chapter both explores a specific moment when Darwin articulated a theory of female mate choice as a mechanism for his theory of sexual selection and highlights many of the issues to which biologists would return in later years.

Darwinian Psychological Continuity between Animals and Mankind

Darwin had grown up with money and with science. His tireless grandfather, Erasmus Darwin, was a member of the Lunar Society and is remembered for (among other things) a posthumously published fanciful poem, *The Temple of Nature; or, the Origin of Society*, in which he described the successive transformations of life from "the first specks of animated earth" to "Newton's eye sublime."[4] Charles's portly father, Robert, trained as a doctor and married one of the Wedgewood daughters. He hoped that young Charles would follow in his footsteps by becoming a medical man. Although Charles never completed his medical training, he did emulate his father by marrying one of his Wedgewood cousins (two of his sisters also married Wedgewood cousins). The intermarriages of the Darwins and Wedgewoods reflected the social status, long-term friendships, and restricted social circles binding these families together.[5]

Charles Darwin enjoyed communicating his scientific ideas to others. He published prolifically throughout his life and maintained a substantial personal correspondence, even though a retiring personality and ill health increasingly kept him close to home. At the age of twenty-nine, he made a literary and scientific name for himself by transforming the field notes and personal journal from his five-year journey on the H.M.S. *Beagle* into an extremely popular travel narrative. Yet he hesitated to publish his naturalistic alternative to the divine creation of all species in their current forms, because of concerns about the reception of such a book among his religious relatives, friends, and peers. Yet, as fate would have it, Darwin was spurred into publishing *Origin of Species* by a letter from Wallace, which outlined a mechanism for species divergence very similar to Darwin's own musings on natural selection. Wallace was fourteen years younger than Darwin and from a working-class family. Although Wallace and Darwin were established correspondents, Darwin held the upper hand scientifically and socially. They agreed to read a joint paper before the Linnaean Society in 1858, but it was Darwin's carefully explicated evidence and ideas in *Origin of Species* that earned him notoriety and adulation.

Deep in the pages of *Origin of Species*, Darwin briefly sketched a second theory that supplemented natural selection.[6] Sexual selection explained why males and females of the same species differed in their appearance and behavior, especially with regard to beautiful mating displays. In comparison with natural selection, Darwin suggested that "sexual selection . . . depends, not on a struggle for existence, but on a struggle between the males for possession of the females; the result is not death to the unsuccessful competitor, but few or no offspring." This struggle for the females might take the form of actual combat, as he deemed common in carnivorous animals, or it might be more peaceful, as was more typical in birds, where physical combat was replaced with male competition for female attention through singing and dancing. Darwin continued by arguing that just as a human animal-breeder could "give elegant carriage and beauty to his bantams . . . I can see no good reason to doubt that female birds, by selecting . . . the most melodious or beautiful males, according to their own standard of beauty, might produce a marked effect."[7] These two mechanisms, male-male competition and female choice, formed the core of Darwin's theory of sexual selection.

In *Descent of Man and Selection in Relation to Sex*, published twelve years later (1871), Darwin enlarged and expanded the scope of sexual selection. He suggested that although natural selection could explain the physical evolution of a single species over time and how that species might splinter and give rise to other, unlike forms, it could not account for the presence of stable variations within a single species, such as sexual or racial variation. He hoped that his theory of sexual selection would provide an explanation for the stable persistence of these intraspecific varieties over long stretches of time. When males competed for access to reproductively available females, any trait that helped them win such struggles, from strength and size to antlers or horns, would be passed along to their male offspring. Only the successful combatants would sire the next generation and, over time, male-male competition would produce selection for larger males and stronger armaments. When females compared a variety of potential mates and chose one of these males to father their offspring, however, weapons were less important to a male's success than was his ability to catch the female's eye. Darwin suggested that, over time, these males would become more beautiful and aesthetically pleasing to garner the attention of the female. Sexual selection couched armaments as a result of instinctual male battles, but male beauty as a result of female aesthetic appreciation. Darwin used male-male competition and female choice, combined, to account for both sexual and racial variation in populations. Sexual selection thereby provided an explanation for the evolutionary origin of these traits, an explanation he thought natural selection unlikely to supply.[8]

In summary, then, whereas natural selection explained variation among species, sexual selection dealt with variation between individuals, male and female, of the same species. Natural selection was based on differential survival, sexual selection on differential reproduction. Natural selection depended on a requisite interaction between individuals of a species and the environment in which they lived, whereas sexual selection was guided by voluntary, perhaps even rational, choice by individuals. In this dichotomous framework, the details of each theory are simplified, but it accurately reflects the framework in which Darwin's contemporaries strove to evaluate and respond to his theories. Noted psychologist and friend of Charles Darwin, George John Romanes, for example, explained the appreciation of beauty in humans as the result of sexual selection. In 1892, he wrote that "this subject matter [evolution] is clearly and sharply divisible into two great classes of facts in organic nature—namely, those of Adaptation and those of Beauty. Darwin's theory of descent explains the former by his doctrine of natural selection, and the latter by his theory of sexual selection."[9] For Romanes and for Darwin, only traits with "life-preserving value" could have evolved as a result of natural selection; sexual selection provided the proper explanatory framework for all aesthetic, sexual traits. When biologists read *Descent of Man*, they contrasted sexual selection, a process of reproduction based on individual aesthetics, with natural selection, a process of survival based partially on environmental conditions.

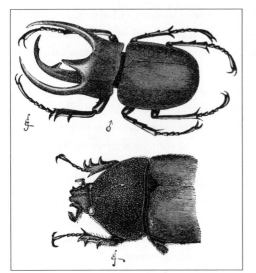 

Sexual differences as the result of male-male competition (left, Atlas beetle, *Chalcosoma atlas*; male above, female below; not to scale) and female choice (right, tufted coquette, *Lophornis ornatus*; female above, male below). Darwin hypothesized that both forms of sexual selection could result in dimorphism (difference in appearance of males and females), but male competition would lead to weapons, such as horns, and female choice would lead to increased male beauty. From Charles Darwin, *Descent of Man and Selection in Relation to Sex*, 2nd rev., augmented ed. (New York: Appleton and Company, 1922), 295, 387.

Darwin strengthened the close association of sexual selection and rational choice through his use of breeding in domesticated animals as an analogy for processes in nature.[10] Over time, a breeder could change his stock's appearance according to his own personal tastes—in other words, subject them to artificial selection. Darwin wrote that "just as man can give beauty, according to his standard of taste, to his male poultry . . . so it appears that female birds in a state of nature, have by a long selection of the more attractive males, added to their beauty or other attractive qualities."[11] Both artificial and sexual selection depended on the vagaries of individual choice—in the former, the selecting tastes of the human breeder; in the latter, the preferences of female animals. Darwin described both artificial selection and sexual selection as resulting from the differential reproduction of individuals in a population. If only a small proportion of the population could breed, then the next generation would disproportionately look like those lucky few.

One of the keys to Darwin's theory of sexual selection was that morphology (color, extravagant feathers) acted together with behavior (courtship dances, mating calls) to create a vibrant display that attracted the attention of females and enticed them into mating. For Darwin, both aspects of this process involved mental processes; on the one hand, females evaluated and chose among males' mating displays; on the other hand, jealous male rivals competed for access to these females. However, Darwin's correspondents and readers focused their reactions primarily on the first of these—female choice. Male-male competition, they thought, could be easily explained as a function of an individual's general vigor and health; only female choice required a mentalistic framework. In such cerebral evaluations lay the problem.

Through his correspondence, Darwin knew that the concept of female choice would arouse skepticism among at least some of his readers. As a preemptive defense against these critics, he carefully outlined exactly what he meant by "choice" in the pages of *Descent of Man* and devoted most of his book to demonstrating that females could "choose." He strove to clarify that, in describing female choice, he did not mean to imply the same kinds of rational deliberation a human would bring to bear when making a similar choice of partner.

> No doubt this implies powers of discrimination and taste on the part of the female which will at first appear extremely improbable; but by the facts to be adduced hereafter, I hope to be able to shew that the females actually have these powers. When, however, it is said that the lower animals have a sense of beauty, it must not be supposed that such sense is comparable with that of a cultivated man, with his multiform and complex associated ideas. A more just comparison would be between the taste for the beautiful in animals, and that in the lowest savages, who admire and deck themselves with any brilliant, glittering, or curious object.[12]

Females, Darwin added, were generally "less eager" than males to mate. A female required courtship as an inducement to mate and exerted her choice by acting "coy" or "endeavouring for a long time to escape from the male." Darwin suggested that although females seemed "comparatively passive," readers should interpret their refusal to mate with just any male as evidence of their discerning sense.[13]

Darwin described female choice as requiring an appreciation of beauty and reflecting an inherent tendency of animals of all kinds to be attracted to showy displays. Females simply needed to determine which male display was the most eye-catching, the most brilliant, the most spectacular. Female choice and male-male competition implied a certain level of mental processing that Darwin hesitated to apply uniformly across the animal kingdom. He believed that the effects of sexual selection, by either female choice or male-male competition, would only be visible in individuals of species that possessed sufficient intellectual capacity to feel the influence of either love or jealousy and to appreciate the beauty of sound, color, and form; he argued that "these powers of the mind manifestly depend on the development of the brain."[14] This hierarchical continuity of emotion and mental function is clearly visible in his arrangement of chapters in *Descent of Man* that elucidate the effects of sexual selection in the animal kingdom. He began his phylogenetic journey with the insects, continued through the evolution of differences in the appearance and behavior of male and female fishes, amphibians, reptiles, birds, and mammals, and ended with a discussion of sexual selection and the evolution of races in humans. In other words, Darwin contended that insects were the "lowest" animal group in which one could expect to detect the effects of sexual selection. Similarly, his comparison of animal attraction and the tendency of "savages" to deck themselves with curious objects revealed his conviction about a continuous and hierarchical nature. All higher animals and humans coexisted on a continuum of behavioral and intellectual complexity, differing in degree rather than kind, ultimately culminating in civilized (male, Caucasian) society.

In fact, throughout *Descent of Man*, Darwin equated the courtship of birds and mammals with human courtship and marriage. All that is required for sexual selection to work, he suggested, "is that choice should be exerted before the parents unite, and it signifies little whether the unions last for life or only for a season."[15] He described the "marriage arrangements" of reptiles and fishes, the "marriage flight" and "marriage ceremony" of butterflies, the "marriage" of birds, and "marriage unions" in quadrupeds.[16] When describing female preference for particular males in birds, he suggested that his readers could only understand the situation if they imagined themselves as aliens come to Earth to view human mating for the first time. "If an inhabitant of another planet were to behold a number of young rustics at a fair, courting and quarrelling over a pretty girl, like birds at one of their places of assemblage, he would be able to infer that she had the power of choice only by observing the eagerness of the wooers to please her, and to display their finery."[17] The aliens' inability to directly ask humans about their mental choices served as an analogy for naturalists' inability to comprehend the internal mental calculations of birds—both required surmising from the available behavioral data. For Darwin, as for many biologists in the coming decades, the topics of human and animal sexual behavior were inextricably intertwined.

Given his belief in the psychological continuity between animals and humans, it is not surprising that Darwin drew parallels between the behavior of men and women and their male and female animal counterparts.[18] He stated that "man is more courageous, pugnacious, and energetic than is woman, and has a more inventive genius." Woman, for her part, surpassed man in her "powers of intuition, of rapid perception, and perhaps of imitation." Darwin could not let such a statement stand unmodified, however, and he added that some "of these faculties are characteristic of the lower races, and therefore of a past and lower state of civilisation." He also quipped that, fortunately, a baby inherits qualities from both parents, "otherwise it is probable that man would have become as superior in mental endowment to woman, as the peacock is in ornamental plumage to the peahen."[19] Sexual selection undermined the place of women in society, painting them as less evolved than men, at the cost of uniting all human races under a single umbrella—mankind.[20]

When he finally turned to the role of marriage choice in humans, in the last chapters of *Descent of Man*, Darwin neatly tied together the two seemingly disparate threads of his book—human evolution and sexual selection—to explain the evolution of human racial types.[21] Earlier in the book he had argued that other attempts to explain the existence of human races in terms of biology—from the idea that races differed because they were adapted to the specific environments in which they lived, to the idea that racial characters were correlated with immunity to disease—all failed.[22] Sexual selection, he continued, was the only viable alternative. Each human race exhibited a unique conception of which human features counted as beautiful; sexual selection according to these disparate standards over many generations would account for the marked differences in skin color, facial features, and degree of hairiness that he considered the defining characteristics of the human races.[23]

Although Darwin sought to explain human ethnic diversity biologically, he also believed that each racial type was dynamic and would continue to change in the future. He admonished his readers to take better care when selecting their mates. Chastising men who gave more thought to the breeding of their horses, cattle, and dogs than to choosing their wives, he suggested they look to the "mental charms and virtues" of their choice and not be attracted to "mere wealth or rank."[24] By doing so, Darwin hoped, men could control both the "bodily constitution" and the "intellectual and moral qualities" of their offspring.[25] In civilized societies, then, marriage choice was reversed from the law of female choice that predominated in the animal kingdom: men chose their wives, and women competed for access to the most eligible men. When it came to extravagant mating displays, women rather than men concocted colorful outfits to garner the attention of the opposite sex.[26] Even so, Darwin recommended that members of both sexes voluntarily refrain from marriage and childbearing if they deemed themselves "in any marked degree inferior in body or mind."[27]

As further evidence for the continuity between animal and human minds, Darwin published *The Expression of the Emotions in Man and Animals* in 1872, close on the heels of *Descent of Man*. In *Expression of the Emotions* he explored a great variety of emotions shared alike by humans and animals, from despair to joy, guilt to astonishment. He also took the opportunity to reinforce some of the points he outlined earlier in *Descent of Man*, including the continuity of two basic emotions central to his theory of sexual selection: love and jealousy. When human lovers met, their love manifested itself physiologically: "their hearts beat quickly, their breathing is hurried, and their faces flush."[28] Subsequent mutual caresses and grooming were common to both animals and humans in the expression of affection.[29] For Darwin, vocalizations in animals not only served to bring members of the same species together during the mating season, but also provided a means for males to "charm or excite the female."[30] He further suggested that music in human societies similarly arose as an expression of emotional arousal: "the habit of uttering musical sounds was first developed, as a means of courtship, in the earliest progenitors of man, and thus became associated with the strongest emotions of which they were capable,—namely, ardent love, rivalry and triumph."[31] Darwin thus emphasized a strong continuity between human and animal vocalizations in the expression of sexual excitement and jealousy, the two emotions crucial to the action of female choice and male-male competition in nature.

In the wake of Darwin's publications outlining the natural origins of humanity and the marked continuity of animal and human with respect to emotion and sexual behavior, many scientists sought to reinforce the animal-human boundary by defining more clearly the essential characteristics that distinguished civilized men from their bestial brethren.[32] Was it their capacity for language, for reason, for aesthetic appreciation, for love? At various times and places, the explanation has relied on each of these. Humans' unique capacity for reason was expressed through their facility with language, through social interactions with other humans—their ability to love, hate, and remember created the cultures in which they lived—or through the divine gift of a soul or intelligence.

Wallace, Divine Intelligence, and the Problem of Choice

Whether or not biologists agreed with Darwin that female choice could act in nature depended mainly on whether their notion of "nature" referred to humans or animals. Many turn-of-the-century biologists, including Alfred Russel Wallace, believed it unlikely that animals were capable of rational comparison and deliberation, and therefore they found it difficult to see how female choice could work in animals. Wallace provides a useful foil to Darwin's attempt to demonstrate the continuity of humans' evolutionary history with the physical, emotional, psychological, and moral foundations of nonhuman animals. Wallace preferred to differentiate human and animal mental and moral capacities. Based on his belief in the unity of man, he argued that in all human races, man exhibited a "wonderfully developed brain, the organ of the mind, which now, even in his lowest examples, raises him far above the highest brutes."[33] For Wallace, this divinely granted mind came to encapsulate the "true grandeur and dignity of man" and allowed humans to escape some of the worst ravages of natural selection through their intellect, sympathy, and moral feelings.[34] Additionally, this view led him to reject sexual selection by female choice in animals, seeing it as an untenable evolutionary mechanism—animals did not possess the ability to reason, thus they could not choose.

In his 1864 address to the Anthropological Society of London, Wallace argued that the races of humankind were formed *before* the "wonderfully developed brain" was independently formed within each race.[35] The inherent differences among the races, he argued, explained why Europeans seemed to inevitably drive to extinction those "low and mentally undeveloped" populations with which they came into contact.[36] Wallace trod a careful line between the polygenists, who insisted on several evolutionary origins for the human races, and the monogenists, who argued for a single evolutionary origin of all humankind.[37] He suggested that although all humans originated from a single progenitor, the complete development of mind, reason, and sympathy took place after the races had already begun to diverge. Thus he encouraged his readers to think of this historical narrative either as a single physical origin for all human races or as a series of independent mental and moral origins of true humanity in each racial type, depending on a reader's proclivity.[38] In all races, he continued, even the least developed human intellect far surpassed that of the smartest animals.

Unlike Darwin, Wallace articulated a spiritually granted division between dumb animals and intelligent humans, suggesting that natural selection could not explain this crucial "origin of sensation or consciousness." He provided three lines of reasoning in support of his claim. First, the brains of all "savage" races were larger than was needed to cope with their current level of civilization. Any variations in brain size among the races were small in comparison with variations in ecological needs among these same populations. He surmised that all humans, including civilized Europeans, possessed a surplus of mental power; a man's brain was an "instrument beyond the needs of its possessor."[39] Much like Darwin, Wallace used his accounts of savage civilizations as proxies for historical accounts of early hominids that he found difficult to re-create from other evidence.[40] As members of these populations did occasionally exhibit great moral and intellectual faculties, Wallace suggested, such abilities lay hidden in the brains of all humans: "the large brain of the savage man is much beyond his actual requirements in the savage state." For Wallace, such advanced mental development, beyond the needs of members of even the higher civilizations—and hence not explicable by natural selection—bespoke the existence of some higher power guiding the development of humans.[41]

Wallace also suggested that the hairy coverings of nonhuman mammals were useful in preventing them from becoming chilled or wet during the course of the day. Yet, humans of all races were relatively hairless, especially down the middle of the back or spine, the body region with the densest hair in other mammals. Wallace could not use natural selection to explain such a gap between the ecological necessities of humans and their physical adaptations. Finally, he described the "singing of savages" as a "more or less monotonous howling." In sharp contrast to Darwin's fascination with human song as evolutionarily continuous with the vocalizations of animals during courtship, Wallace argued that sexual selection could not explain such raucous noise. In conclusion, Wallace stated that just as humans can influence the development of animals and plants in agriculture, "a superior intelligence has guided the development of man in a definite direction, and for a special purpose."[42] Wallace was to maintain this position for the rest of his life, eventually suggesting that the concurrent development of intelligence and morality in humans comprised only the most recent transition in life that could not be explained by natural selection, being preceded evolutionarily by the transitions from inorganic chemicals to organic life, and from plant (insensitive) to animal (sensitive) life.[43]

Consistent with this view, Wallace believed that female animals were not capable of evaluating the aesthetic appeal of males' courtship displays and choosing a mate based on their impressions. In an 1877 paper on the colors of plants and animals, Wallace followed Darwin by demarcating female choice as "voluntary" sexual selection (as a foil for males' struggles with other males for possession of the females). Wallace added a footnote clarifying his meaning and suggesting that Darwin's term "voluntary" was not strictly applicable and proposed replacing it with either "conscious" or "perceptive" sexual selection. Both terms, he suggested, were free from ambiguity and made no difference to the overall argument.[44] Wallace suggested that only one factor supported Darwin's theory of "perceptive" sexual selection: most likely, male persistence eventually won over the naturally "coy" females. Whereas males were likely to accept any female, females typically accepted a mate only after his persistent and energetic display.[45] However, Wallace doubted that such conscious choices could explain natural beauty in the animal kingdom.

Wallace provided an alternative explanation for the origins of beauty in nature by arguing that male extravagance was normal. Bright colors and oddly shaped feathers or horns were a body's natural expression of extreme vigor. Wallace believed this explained why displaying male birds had puffed-out, conspicuously colored chests—the color was due to the excess energy of the bird's singing. Further, in Victorian England, males were known to be more vigorous than females, and thus it only made sense that males would be more colorful than females. If any extra explanation was needed, it was to show why females were so dull in color. Wallace explained this with natural selection: females were under greater evolutionary pressures to be dull. If a predator caught a female during the nesting season, she lost not only her life but the lives of all future offspring as well.[46] Wallace's disenchantment with sexual selection in the animal kingdom reflected sentiments shared by many of his peers.

Far more controversial were Wallace's views on how selection might act in humans, not according to natural selection (which he had dismissed) but through the very capacity granted man by a spiritual power—his mind. Conscious selection of mates, Wallace reasoned, might offer a way for humans to control their evolutionary future. However, social rules and mores would have to change first.[47]

Yet Wallace worried that neither beauty nor intelligence correlated with class, and people's choice of a marriage partner was often dictated by financial concerns.[48] He further added that, given the new research on heredity in humans, education seemed an unlikely solution to this dilemma, as parents could pass along to their offspring neither a commitment to self-improvement nor their accumulated wisdom. Education and habit were not heritable human traits. The evolutionary future of human society might appear grim in the light of these facts, but Wallace was an optimist.

He argued that some form of biological selection was required to further the "improvement of our race." What form would this selection take? Wallace rejected a proposal suggesting that only a few individuals in each generation should be allowed to reproduce; he reacted in horror to another proposition to abolish marriage completely and allow women to bear children with temporary husbands of good quality. In both cases, he objected to legal intervention by the state in determining the right of individuals to marry and bear children. Instead, Wallace suggested that when people were "free to follow their best impulses," they would spontaneously create a "system of selection" that would "steadily tend to eliminate the lower and more degraded types of man, and thus continuously raise the average standard of the race."[49] Economic equality and universal education would free women and men from financial concerns, leaving them to follow their natural instincts. He believed that the first step in social progress was to lift persons of lower social status (including men of the lower classes and all women) to equal economic opportunity. Only then would evolution through sexual selection be able to act effectively.

It may be taken as certain, therefore, that when women are economically and socially free to choose, numbers of the worst men among all classes who now readily obtain wives *will be almost universally rejected . . .* it is the one brilliant ray of hope for humanity that, just as we advance in the reform of our present cruel and disastrous social system, we shall set free a power of selection in marriage that will steadily and certainly improve the character, as well as the strength and the beauty, of our race.[50]

For men, in whom "the passion of love is more general, and usually stronger," the sole outlet for their sexual desire would be through marriage. As a result, every woman would receive multiple offers of marriage and, as was their nature, women would be extremely careful in their selection of a husband. In fact, released from the burdens of their financial future, Wallace hoped, some women would choose to remain single rather than settle for a "diseased," "weak," "idle," or "selfish" husband. The effect of this sexual selection would be to eliminate the most unfit members of society. The "cultivated minds and pure instincts of the Women of the Future" would slowly and steadily improve the race without any legal intervention.[51]

Wallace suggested that female choice could act on human evolution with greater effect than on animal evolution because the artificial conditions of humans' urban life differed significantly from the natural ecological habitats of animals. As a result, in humans, the evolutionary effects of female choice were not swamped by the effects of environmental conditions on individual survival, as they were in animals. "Sexual selection," Wallace wrote, "possesses the potentiality of acting in the future so as to ensure intellectual and moral progress, and thus elevate the race to whatever degree of civilization and well-being it is capable of reaching in earth-life."[52] Through economic equality, Wallace hoped, female choice, acting through marriage selection, would eventually become an even greater force of evolutionary change in society.

Wallace's views on sexual selection in animals and humans stemmed from his initial commitment to female choice as a conscious mental activity, akin to artificial selection. His belief in a fundamental divide between human rationality and animal instincts led him to reject notions of female choice in animals at the same time that he advocated sexual selection in humans as the key to our evolutionary future.

The Reception of Female Choice in Humans and Animals

After Darwin's death, the concept of sexual selection began to lose its coherence, as biologists and psychologists struggled to frame female choice without recourse to assumptions about the internal machinations and aesthetic preferences of animal minds.[53] Darwin's cousin and leading eugenicist Francis Galton, for example, suggested that female choice was a process by which the sexes formed sympathetic matches—tall women tended to marry tall men, intelligent men often sought intelligent wives, a kind of assortative mating.[54] Karl Pearson, noted mathematician and statistician, distinguished between two types of female choice: assortative mating (following Galton) and true female preference (which he called "tribal taste"). In both cases, Pearson stressed that a female's choice was relative; she would not reject all males according to an absolute standard but would instead mate with the most pleasing male she encountered.[55] First-wave feminists easily appropriated such theories of mate choice as platforms for reforming British and American courtship habits and systems of marriage. Yet, such conceptual diversity, coupled with increasing concern about the mental calculations required for animals to choose their mates based on aesthetic preferences, spelled difficulty for Darwin's theory of sexual selection in animals.

For Darwin's and Wallace's readers, female choice as a mechanism for evolutionary change continued to raise hoary issues about the biological definitions of masculinity, femininity, racial distinctions, and rationality as crucial for distinguishing humans from other animals. Several feminists highlighted Darwin's caesura between human and animal marriage choice: whereas female animals chose their mates, women were at the mercy of male choice in Victorian England. Such an unnatural state of affairs in modern society could only lead to degeneration, they argued. Other feminists argued that Darwin's theory simply reiterated what everyone already knew—that the sexes were physiologically and socially divided into two separate but equal social and biological realms, where women proudly bore the moral and aesthetic standards of the race. Interestingly, few feminists objected to Darwin's framing of female and male sexuality as oppositionally related—female "coyness" and male "eagerness" would not be systematically called into question until much later in the twentieth century. The implications of sexual selection for human society captured the imagination of the larger scientific community and educated public, even if sexual selection in nonhuman animals did not.

Following Darwin's publications, especially *Expression of the Emotions*, a few biologists sought to systematically explore the psychology of animals. Certainly, Darwin's friend (some have called him a disciple) George Romanes had this goal in mind when he wrote *Animal Intelligence* and *Mental Evolution in Animals* in the early 1880s.[56] Romanes tried to establish the continuity of mental abilities in animals and humans; the roots of all aspects of human mental abilities, he argued, could be found in the behavior of nonhuman animals. Human minds, however, were qualitatively more complex in their ability to both perceive and respond to the changing world around them. In *Mental Evolution in Animals*, Romanes posited "choice" as the objective criterion by which he would distinguish whether or not animals possessed a mind: "agents that are able to *choose* their actions are agents that are able to *feel* the stimuli which determine the choice."[57] Romanes willingly attributed an aesthetic "emotions of the beautiful" to all animals exhibiting colorful secondary sexual characters (the result of sexual selection), right down to insects, spiders, and crabs.[58] Choice, for Romanes, was intimately tied to feeling.

Other psychologists found Romanes's link between choice and feeling rather problematic. British psychologist and sexologist Patrick Geddes and his friend and student J. Arthur Thompson, for example, co-published *The Evolution of Sex* to great acclaim in 1889.[59] Their approach to female choice and sexual selection highlighted several crucial transitions in the continuous evolutionary history of mate selection, from animals to humans—notably, a switch from an emotional to an intellectual basis of choice.[60] Geddes and Thompson argued that in primitive animals, hunger (the nutritive drive) and love (the reproductive drive) were self-serving, easily gratified instinctual behaviors. Over the course of evolution, in higher animals these basic instincts were refined, through individuals' increased awareness of their community and self, into social instincts. In an ideal community, social cohesion was maintained through altruism and an appreciation of the common good. More specifically, Geddes and Thompson suggested that the process of choosing a mate was less emotional and more intellectual in humans than in animals, less physical and more mental. Furthermore, humanity had attained its current evolutionary position through the complementary sexual roles of active men and passive women, visible at even the earliest stages of life (compare small and vigorous sperm with large and ponderous eggs).[61]

A few early feminists immediately picked up on both Darwin's description of "reversed" sexual selection and Geddes and Thompson's physiological/psychological evolutionary model, and argued that the place of women in Victorian culture was "unnatural."[62] The evolutionary future of civilized society as they knew it would be lost to the lower races unless the situation were remedied. If women were not given free choice of husbands, the evolutionary future of upper-class society would be jeopardized. And, to exercise free choice in husbands, women needed to be freed from the social and economic chains that bound them—otherwise, most women would be forced to marry for money rather than freely choosing the man they loved.[63]

An American suffragist from Michigan, Eliza Burt Gamble, published *The Evolution of Woman: An Inquiry into the Dogma of Her Inferiority to Man* in 1894. Gamble argued that despite popular conviction, women were morally and aesthetically superior to men because of thousands of generations of selecting the "best" males with whom to mate.[64] Female choice, to Gamble, not only had honed the intellectual capacities of females but also had driven males to excel in all their endeavors in order to attract a female mate. She believed that men were naturally selfish and egotistical (as sexual selection dictated that their biological function was to look beautiful and drive other males away), while women were maternal and selfless. The maternal instinct in females, while ultimately derived from natural selection, was amplified by sexual selection and became the basis for all social sentiments.[65] To be capable of choosing the best male, Gamble argued, women must possess a keener aesthetic sense and intuitive intellect than men. She also contended that inasmuch as men were superior, it was only because the woman "made the male beautiful that she might endure his caresses." Furthermore, Gamble accepted Darwin's suggestion that the sex roles normally defined by sexual selection were inverted in Victorian society. Society was thus governed by egotistical (male) rather than altruistic (female) tendencies. Gamble saw the lack of female involvement in mate choice as detrimental to the continued vitality of society. Women's inability to freely choose a mate, she argued, would eventually lead to evolutionary stagnation and social inferiority; "mankind will never advance to a higher plane of thinking and living until the restrictions upon the liberties of the female have been entirely removed."[66] In her conclusion, Gamble asserted that the history of humankind was "a series of alternating periods of development and decay."[67] Each period of decay corresponded to a time when women were ostracized from society and repressed. Gamble further argued that because of their very high birthrate, the lower races would inevitably overcome more advanced society—unless woman could regain her rightful place.

Charlotte Perkins Gilman also saw the saving grace of modern society in female choice. Gilman was a prolific author, producing more than a dozen books of nonfiction as well as fiction, and a prominent social critic in the United States from the 1890s until her death in 1935.[68] Gilman is perhaps best remembered today for her fictional stories, *The Yellow Wallpaper* (1899) and *Herland* (1915). Revisiting Gamble's sentiments on the "unnatural" position of female choice in Victorian society, Gilman published *Women and Economics* in 1898.[69] Here, she argued that women's economic independence from men would allow women the freedom to exercise their intuitive choice in partners, thereby unfettering the natural processes of social evolution and restoring the normal balance of the sexes.[70] She emphasized that the unnatural social position of women as perpetual consumers prevented them from producing anything of social value, and she viewed most women as social parasites, producing an overall loss of social energy. Without immediately returning women to their rightful place of power, Gilman argued, society would stagnate, hindering and perverting economic development.[71]

Frances Swiney, a British feminist and social purity activist, also believed that moral and social truth could be found in nature. She sought to understand the laws of nature, including sexual selection, so as to construct a better future for humanity.[72] First published in 1899, Swiney's *The Awakening of Women or Woman's Part in Evolution* called for the inclusion of women in the workforce.[73] Building on Geddes and Thompson, Swiney used evolutionary and other scientific reasoning to argue that women were anabolic (keepers of the "industrial arts of peace, the virtues of the home and of the family, and the ultimate welfare and happiness of the state"), while men were catabolic (excelling in invention, science, art, beauty, and knowledge).[74] As a result of evolutionary selection, however, women were superior to men and would control the evolutionary future of humanity through their will. She claimed that women had been created superior to men, as could be seen in the power of female choice in animals. Over time, women of the lower races had degraded themselves by pandering to male sexual passions. Only by guarding social morality could women atone for their fatal mistake in the Garden of Eden and stop the evolutionary retardation of their race and nation.[75]

These feminists, who focused on the application of sexual selection theory to humans, remained mostly silent on the implications of choice in animal evolution. Their theories instead formed part of a much larger discourse on the "woman question" in the late nineteenth and early twentieth centuries.[76] For example, both Gilman and Swiney argued that females were clearly superior to males because of their innate sexual differences—women's capacity to reproduce and their control over mate choice.[77] Both sought to liberate women through female choice, and they did so by reinforcing constructions of gender that emphasized a woman's role in society as the selecting sex.[78]

In their discussions of sexual selection and human society, most feminists steered clear of discussing animal psychology, perhaps as an attempt to distance themselves from charges of mawkish sentimentality. Despite Romanes's desire to place animal psychology on a solid scientific footing, the increasing popularity of nature study and the rise of pet culture in middle-class Britain and America brought animals into domestic spaces with greater frequency, strengthening the ideological connections between anthropomorphism and emotional concern for animals.[79] Explanations of natural history to children, often with explicit moralistic lessons embedded in stories of backyard animal antics, elicited scientific concern with "nature fakery."[80] The association between female sentimentality and animal welfare extended well beyond children's education, as women were also deeply involved in the antivivisectionist movement and the burgeoning Audubon Society for the protection of birds. First-wave feminists with a corrective eye to evolutionary theory, therefore, would have found it advantageous to avoid discussing animal behavior and concentrate on their primary goal, the future of human social evolution.

Most scientists, however, remained more concerned with mentalistic assumptions about nonhuman animal minds than with challenging Darwin's suggestion that *female* animals (rather than their male counterparts) exhibited control over mate choice. For example, when a young woman at the University of Michigan, Cora Daisy Reeves, investigated the courtship habits of the rainbow darter, she concluded in a 1907 paper that there was no reason to attribute the male courtship displays or bright colors to choice on the part of the females. In line with the biological attitudes at the time, Reeves instead attributed a stimulatory function to the male's vigorous pursuit and repeated tapping of the female's body. She earned her Ph.D. in 1917 and published at least two other papers, on the ability of fish and rats to discriminate different wavelengths of light.[81] Her later research on behavioral stimuli rather than choice reflected a growing tendency in American and British research on animal psychology to see animals as living, controllable machines.

One conspicuous manifestation of this new trend was London-born psychologist Conway Lloyd Morgan's "canon" suggesting that psychologists should not assume the existence of higher mental states than were needed to explain animal behavior.[82] In other words, animal psychologists should not call behaviors "choices" unless they had good evidence that the animals were actually "choosing." Morgan spent a good part of his book, *Animal Life and Intelligence*, critiquing Darwin's theory of mate choice.[83] Morgan was skeptical that a "consciously aesthetic motive" in animals (even birds) could explain the origin of beauty in birds or other animals. The appreciation and judgment of beauty, he suggested, was a matter of human rather than animal psychology. Later in the book, Morgan clarified his position. Animals possessed "perceptual volition," whereas humans were capable of "conceptual volition." In other words, given a set of circumstances, animals might choose between two possibilities with which they were presented (left or right, male 1 or male 2, blue or red), yet they could not reflect on their choice, nor imagine that, under altered circumstances, they might have chosen differently. Humans alone possessed the capacity for self-reflection and a conscious appreciation of their choices, including aesthetic appreciation, which Morgan linked to conceptual thought.[84] Morgan also agreed with Wallace (as a way of disagreeing with Darwin and Romanes) that human intelligence and consciousness could not be explained by the "law of elimination," natural selection.[85] Morgan and his canon became icons both among the animal psychologists who founded American behaviorism in the early twentieth century and among the European zoologists whose ethological theories of animal behavior became popular in mid-century.[86]

In the last decades of the nineteenth century, Darwin's theory of sexual selection by female choice educed rocky debates over the value of female choice as a tool for understanding beauty in the animal kingdom and marriage choice in human society. For many of these scientists and feminists, as for poets and philosophers before them, courtship and sexual behavior in humans were associated with beastly passions of the flesh, rather than with rational consideration. Female choice posed an unexpected conundrum, in which Darwin ascribed to animals—that is, to female animals about to engage in sexual activity—the mental advantages of aesthetic appreciation and conscious choice. No wonder this concept met with such stark resistance among biologists who, like Wallace, claimed that female choice could never be demonstrated in animal nature.

It would be good to remind ourselves that the turn of the century was a difficult time for evolutionary theory in general.[87] For many experimental laboratory biologists, selection (either natural or sexual) did not seem to be required to understand how species might have changed over time. The rediscovery of Gregor Mendel's theory of hereditary factors seemed to offer a viable alternative explanation of species change, just as Jacques Loeb's mechanistic model of animal bodies and minds provided a framework for studying an animal's behavior without assuming access to its inner life or emotional experiences.[88]

The lasting legacy of the debates between Darwin and Wallace over the legitimacy of female choice in animals and people was that only humans could choose and only humans had aesthetic preferences, although this was not because biologists found Wallace's arguments more persuasive. As we will see in the following chapters, the scientific study of animal courtship behavior in the early twentieth century both provided lessons for the evolutionary future of British and American societies and reified a cognitive-aesthetic divide between human and animal. As biologists framed the evolutionary future of humanity in mathematical terms, their formulas reflected the social importance of reproduction as the key to greater intelligence and beauty in future populations. Although biologists lauded female mate choice as crucial to human evolution, they continued to deny that animals were endowed with sufficient reason to choose one mate among many.

CHAPTER 2

Progressive Desire

Rational Evolution after the Great War

The varied and individualized love-making of humanity differs fundamentally from the courtship of birds. The human lover woos with the cerebral cortex, he (or she) is plastic and responsive, and adapts the means to the occasion. The impassioned bird woos ardently but automatically with the corpus striatum. In one case the individual of the opposite sex is a problem; in the other an exciting stimulation. The human lover may do a thousand things; the courting bird is an elegant determinate machine.
— H. G. Wells, Julian S. Huxley, and G. P. Wells,
The Science of Life, 1931

In the early decades of the twentieth century, two different ways of describing animal courtship allowed American and British biologists to generalize between animal and human reproductive behavior. The first was to couch animal courtship as an evolutionary forerunner to marriage choice in humans, within the framework of Darwin's theory of sexual selection. Within a eugenic framework, this narrative of evolutionary progress or degeneration served as one of many mechanisms differentiating some human cultures from their bestial ancestors and zoomorphizing other cultures as intellectually and morally no more advanced than common animals. The second was to portray animal courtship as a physiological process akin to each act of human "love-making" (to borrow a euphemism from Julian Huxley). This direct equation of human and animal copulation provided an alternative to the hierarchical relationships inherent to early twentieth-century narratives of evolutionary love. Whereas mate choice in sexual selection implied comparison and aesthetic appreciation on the part of the chooser, a physiological description of courtship painted animal and human (re)actions as nonrational and the result of emotional stimulation. Biologists applied both models of courtship to human sexual relations, but continued to struggle with the question of whether female choice could be a valid mechanism of evolutionary change in animals.

In the 1900s through 1930s, professional zoologists almost universally rejected notions of mate choice in nonhuman animals.[1] This rejection resulted partly from the rise of gonadal (hormonal) theories of sex development, which seemed to obviate the need for an evolutionary explanation of the same phenomenon. Far more damning in the eyes of zoologists, however, were the assumptions about animal psychology necessary to accept the idea of choice-based behaviors in nonhuman animals. For female animals to choose their mates based on a courtship display, it seemed they would need to evaluate the relative aesthetics of the males' displays and make a reasoned decision based on that preference.

Zoologists salvaged Darwin's theory of sexual selection by circumscribing or reconceiving what he originally intended by "female choice." True choice they restricted to humans; although nonhuman animals might not be capable of choosing their mates in such a fashion, humans clearly *were* capable of such deliberations, and proper marriage choice in humans could lead to smarter and more beautiful children in future generations. Additionally, apparent mate choice in animals need not imply mental deliberations but could just as easily result from differences in the emotional stimulus provided by a mating display: male animals whose courtship displays were more effective at stimulating females would be more likely to mate and more likely to stay with their mate for the breeding season. Some zoologists also applied this notion of ritualized, emotional courtship to humans, by analogizing animal courtship to human foreplay (both leading to coition), the function of which was to keep a couple happily united after the wedding.

These modifications to the theory of sexual selection involved an *emotional continuity* in the function of human and animal courtship and simultaneously served to reinforce a *mental divide* between animal and human. True female choice was a rational (or at least semi-rational) process that zoologists could safely attribute to humans only. Within a Darwinian framework, zoologists in the early twentieth century continued to define a bright line between uncivilized animals and civilized human societies.[2]

Popular discussions of sexual selection centered on the role of female choice as a mechanism of human evolution. Natural selection might still be affecting "uncivilized" human societies, and certainly still affected animal species, where individuals struggled merely to stay alive long enough to reach reproductive maturity. But to some zoologists it seemed that in modern human societies, natural selection based on differential survival could no longer act as an efficient evolutionary mechanism; technological and medical advances prevented the uninhibited effects of environmental factors on the survival of individual humans. If any phenomenon could affect the evolutionary future of modern societies, differential reproduction was considered a more likely candidate than natural selection. Members of the social hygiene movement, therefore, sought to educate people on how to make superior reproductive choices and encouraged "well-born" families to have more children.[3] This form of human engineering resonated strongly with the rationality of female choice in sexual selection.[4] Through proper female choice, sexual selection could act to improve the genetic stock of humans over time, making women more beautiful and men more intelligent.

Further, some Britons and Americans may have found comfort in the thought that rational breeding of humans was a kinder approach to social evolution than the *Allmacht*, or all-sufficiency, of natural selection ostensibly advanced by the Germans around the time of the First World War. For Vernon Kellogg, an American zoologist and professor at Stanford University, Germans' aggression arose from their insistence that natural selection could act as a guide for understanding the persistence of empires in civilized cultures. Kellogg's *Headquarters Nights*, published in 1917, suggested that German military advisors believed that "bitter, ruthless struggle" between different human groups "not only must go on, for that is the natural law, but it should go on, so that this natural law may work out in its cruel, inevitable way the salvation of the human species."[5] In effect, Kellogg contended, Germans instigated the Great War as a test of their evolutionary status in the world. Although Kellogg's book should certainly not be read as an accurate reflection of German attitudes toward evolution, it was great propaganda.[6] He hoped to convey to his readers a moral lesson: unqualified support of natural selection as the sole guiding force of social evolution could lead to untempered aggression and war. In comparison with natural selection's emphasis on survival of the fittest, sexual selection through female choice offered a rational, peaceful, choice-based approach to understanding and controlling the evolutionary future of society.

Between 1900 and 1939, the pages of the *New York Times* bore witness to sustained popular interest in human female choice and sexual selection.[7] Several articles quoted William Jennings Bryan as saying that sexual selection had been "laughed out of the classroom," and Henry Fairfield Osborn pointed to sexual selection as a "debatable" aspect of evolutionary theory. Although these articles painted sexual selection in a less than positive light in terms of animal evolution, they did not discuss female choice as applied to the evolutionary future of humans. When newspaper articles did invoke sexual selection as an evolutionary force in humans, they contended that if women made good choices in marriage partners, then the beauty and intelligence of humans would increase in future generations. In his articles for the *New York Times*, George Bernard Shaw repeatedly emphasized the importance of letting women make their own choice in husbands, without economic or social class restrictions, allowing true "female choice" to operate in society. An article published in 1921 went so far as to argue that London's rat problem could be solved if exterminators would use sexual selection instead of rat poison.[8] According to the headline, exterminators should "KILL ONLY THE FEMALES, Then the Males Will Fight Each Other and Eventually Destroy the Breed." The natural aggression of males, if left unchecked, would prove the ruin of the rat population, just as it would for humanity. Rational selection by females, on the other hand, could save the species from extinction. By the late 1930s, however, the onset of the Second World War directed popular attention away from female choice and toward more pressing matters. Sexual selection and female choice did not reappear in the pages of the *New York Times* until the 1970s.

The Problem with Beauty and Aesthetics in Animals

In the early twentieth century, few biologists were satisfied with either Wallace's application of natural selection or Darwin's use of sexual selection as an explanation of differences between male and female animals. Simultaneously, mutually reinforcing trends in biological studies of animal behavior served to downplay female choice as a valid mechanism of evolutionary change in animals. Zoologists interested in animal behavior began to create new theoretical models of reproductive behavior and new vocabularies to describe behavior that avoided animals' aesthetic judgment and capacity to reason. Experiments on mating behavior in fruit flies and moths seemed to demonstrate that insects did not possess the mental apparatus necessary for choosing a mate. Additionally, the traits that female choice seemed to explain—sex differences in behavior and appearance (sexual dimorphism)—could be explained by other factors. Experimental evidence from physiologists, for example, showed that physical differences between the sexes could be explained hormonally or developmentally rather than evolutionarily. The remarkable dances and displays of male birds during the mating season could be attributed to the necessity for communication or territorial defense. Zoologists simply did not *need* sexual selection to explain such sex differences. Further, as a function of mapping the evolution of social and sexual behavior hierarchically, studies of animal behavior erected an increasingly strong divide between animal and human ability to reason. Humans represented the pinnacle of behavioral complexity. In constructing these comparisons, biologists reinforced the traditional wisdom that animals and humans differed fundamentally in their intellectual capacities. Whereas humans were capable of language and rational decision-making, animals were not.[9] Such divides between beastly passions and human rationality can be traced from Aristotle and Plato through the writings of Descartes in the seventeenth century and Buffon and Linnaeus in the eighteenth century. In the late nineteenth century, Darwin and Romanes had once again called into question a strict dichotomy between animal and human, a division that was subsequently reinforced by their detractors.[10]

One of the most frequently cited summaries of pro- and anti-Darwinian sentiments in zoology at the turn of the century was Vernon Kellogg's *Darwinism To-day*, published in 1907; Kellogg's outlook on sexual selection is best described as optimistically defensive.[11] He spent twenty-two pages of the volume documenting the evidence and criticisms that biologists had mounted against sexual selection, and slightly less than two pages in defense. Of the criticisms he detailed, all were directed at female choice rather than male-male competition. Kellogg suggested that the evolution of defensive and offensive weapons in males could be explained by natural selection alone. Female choice, however, could not be collapsed into natural selection and posed a serious stumbling block for zoologists.

Kellogg closely followed Darwin's arguments in *Descent of Man* when he argued that sexual selection was limited in its potential efficacy to those species in which female choice could act to reduce the total number of offspring produced by some males in the population. This could happen either when females were in short supply (male-skewed sex ratio) or when some males mated with many females (polygamy), in both cases leaving an unlucky portion of males with no mate at all.[12] Not all species with dramatic mating displays or coloration in males fulfilled these criteria, however, making it seem unlikely that sexual selection was the correct explanation. Further, in species with equal numbers of males and females, there was no evidence of aesthetic development other than the dramatic displays of males during courtship, and no evolutionary justification for the development or presence of the ability to choose.[13]

Even more damningly, Kellogg noted that most evidence seemed to point away from animals' ability to choose. Zoologists had shown that females were either passive, mating with the first male who wandered by, or when given the option of several differently colored males, failed to demonstrate a consistent preference. Kellogg devoted several pages of *Darwinism To-day* to describing the experiments that A. G. Mayer, president of the Cambridge Entomological Club (in Cambridge, Massachusetts), conducted on moths.[14] Mayer scraped off all the colorful scales of male bombycine moth wings and then tracked whether these males were capable of obtaining a mate. They were. Next, he clipped off male wings, glued on female wings, then looked at whether those males would be able to attract a mate. Yes, again. Finally, he clipped off female wings and replaced them with male wings, to see whether bright colors on females would prevent males from mating with them. These females, too, obtained mates. Mayer concluded that a female attracted males by emitting a scent that the males used to find her in the dark. Despite their bright coloration, the visual appearance of the moths could not be a factor in mate attraction, and females certainly did not exercise any choice over male suitors based on their coloration.[15] To biologists in the early twentieth century, such experimental evidence indicated that mate choice, at least in moths, was highly unlikely. The authors of a 1931 textbook, *The Science of Life*, noted, without attribution, that "we have to realize that the male moth has no idea of a mate; he merely possesses mating reactions; and these are fired off by one stimulus and one only—a particular smell."[16] Mayer's article was cited numerous times over the next three decades as evidence refuting Darwin's theory of sexual selection.[17]

Even before his research on *Drosophila* made him famous in biological circles, Thomas Hunt Morgan objected to the notion of sexual selection because of the lack of any explanation for, or evidence of, female preference in animals—even the chickens he bred during the summer at Woods Hole Biological Laboratory.[18]

Shall we assume that still another process of selection is going on, as a result of which those females are selected by the males that appreciate their unusual beauty, or that those females whose tastes have soared a little higher than that of the average (a variation of this sort having appeared) select males to correspond, and thus the two continue heaping up the ornaments on one side and the appreciation of those ornaments on the other? No doubt an interesting fiction could be built up along these lines, but would anyone believe it, and if he did, could he prove it?[19]

Following his doctoral research under Morgan's supervision at Columbia University, Alfred Henry Sturtevant conducted studies in which he gave female fruit flies their choice of males, and males their choice of females.[20] Sturtevant concluded that "in *Drosophila*, neither sex exercises any 'choice' in the selection of a mate. A female that is ready to mate will accept any male, and a male that is ready to mate will do so with the first female that will allow him to."[21] If a male had a mating advantage over another male, this advantage was due to his innate vigor—in other words, natural selection, not sexual selection. What had been interpreted as mating choice in the past probably was not, Sturtevant argued. Females, naturally reticent, needed stimulation to persuade them to mate. Once sufficiently stimulated, a female would mate with the first male she encountered. Sturtevant's experiments seemed to indicate that the aesthetic comparisons necessary for female choice were beyond fruit flies' mental ability.

Physiologists also discounted the role of female choice as an explanation for male beauty in animals.[22] They regarded evolutionary causes of sex differences as mere speculations and instead relied on gonadal (hormonal) explanations of male-female differences.[23] Male and female gonadal secretions differed in their developmental effects on male and female bodies, producing (in theory) two distinct adult sexes. During the mating season, the extra-excitability of organisms naturally led to increased hormone production and therefore to seasonally extravagant phenotypes. Even genetically inclined Thomas Hunt Morgan adopted this physiological line of reasoning by 1932. If one injected sex hormones into birds, Morgan reasoned, the birds' appearance and behavior would change so that individuals mimicked the opposite sex. The chromosomal sex would be overridden by the hormonal sex; therefore, sexual selection was not needed and should not be invoked to explain sexual dimorphism.[24]

In the light of these experiments, Kellogg doubted that evidence of female choice would ever be produced, especially in insects. He offered merely two lines of possible defense. First, however ineffectual the theory might appear, he suggested, it should be maintained as an explanation of sexual dimorphism until a better explanation could be found. Second, the mental and aesthetic connotations of female choice could be avoided if zoologists framed female behavioral responses to mating displays as emotional responses to stimulation. Following the lead of Conway Lloyd Morgan, Kellogg argued that "this so-called choice is one of impulse, not deliberation: it is an imperative reaction to a sufficient stimulus . . . 'it is a perceptual choice arising from impulse rather than an ideational choice due to motive and volition.'" Courtship was required to generate in the female "an impulse sufficiently strong to overcome her instinctive coyness and reluctance." The variability of male mating success on which Darwin's theory of sexual selection then rested was the result not of aesthetic preference in females, but of the "effectiveness of the [male] ornament, or call, or behavior."[25] This reframing of female choice in nonrational terms was especially timely, given the sentimental inaccuracies attributed to animal cognition by popular writers, who were often disparaged by practicing scientists as misrepresenting the brutality of the struggle for survival in the natural world—as "nature fakers."[26]

Evidence of female choice in insects did exist in 1907, but Kellogg did not mention any of this experimental or observational support for sexual selection in animals, perhaps because such support was found largely among amateur enthusiasts rather than among professional biologists—George Peckham and Elizabeth Peckham's experiments on mating behavior in spiders, William Plane Pycraft's theories of animal behavior, and Edmund Selous's careful observations of bird courtship.

In their publications on the evolution of courtship rituals in spiders, Peckham and Peckham attempted to provide evidence arbitrating the long-standing dispute between Darwin's original formulation of sexual selection and Wallace's critiques of the theory.[27] They argued that Wallace was incorrect: extravagant coloration could not be a simple physiological expression of vigor. Although male spiders were more beautifully colored than females, the females were far more active. Additionally, the spiders with the most spectacular coloration were the orb-weavers, which simply sat on their webs and waited for prey; in comparison, the more active jumping spiders were drab. Further, ornamented male spiders displayed only when females were present and ensured that their ornaments were in full view of the females at all times during courtship. The Peckhams concluded that the coloration and tufts of hair in male spiders must be the result of Darwinian sexual selection. Their results seemed to disprove the validity of Wallace's explanation of vigor as the cause of sexual dimorphism in animals, but they left untouched Wallace's other critique of Darwin's theory, that nonhuman animals lacked the mental development to evaluate and choose a mate.

William Pycraft, a well-known, if not universally well-regarded, indoor zoologist at the British Museum of Natural History, published *The Courtship of Animals* in 1913, also arguing on behalf of a modified form of sexual selection.[28] Following from Kellogg's earlier suggestion, Pycraft postulated that male display and ornament did not necessarily have to act on a female's aesthetic sense, but could emotionally excite her instead. Male display could be seen as raising the female to a state of "sexual exaltation" where she would be ready to mate. His chapter summaries included such topics as "happiness and hartebeestes," "birds of paradise in the toils of love," "the frog that would a-wooing go," and "spiders and conjugal bliss."[29] Pycraft's flowery language, however, landed him in trouble with other biologists. For example, when a curator at the American Museum of Natural History in New York was asked about Pycraft's book, he replied: "Unfortunately the book you mention is extremely popular and I have been unable to secure much of real value from it."[30]

Perhaps the most lasting support for Darwin's theory of sexual selection came from Cambridge-educated naturalist Edmund Selous.[31] Selous published prolifically on animal behavior throughout the first three decades of the twentieth century, including studies on courtship in birds.[32] After hundreds of hours of observing wild birds, Selous argued that female choice *must* occur in nonhuman animals. The females he observed did not respond equally to all male displays, and he noted that some males obtained many mates and others never found a mate at all. Selous also described male-male competition as a kind of substitute "war dance." Rather than actually fighting, he reasoned, males displayed to other males as a way to maintain their own territories. Selous's views on animal mating did not gain traction among other naturalists. He attributed his peers' distaste for his theories to an unlikely, but suggestive, source. "I have never been able to follow the arguments used against sexual selection," he began, "but only the prejudice behind them which is called up more thoroughly, I think, in this emotional region than in anything physiological merely or even intellectual . . . [Biologists] can stand an intrusion of the animal as a pre-human merely, but not as a pre-lover."[33] Only one influential researcher took Selous's ideas about sexual selection to heart—a young man named Julian Sorell Huxley.

Mutual Choice, *Married Love,* and the Great Crested Grebe

Julian Huxley was the grandson of Thomas Henry Huxley, "Darwin's Bulldog," and older brother of Aldous Huxley, the author of *Chrome Yellow* (1921) and *Brave New World* (1932). An avid birdwatcher since childhood, Julian Huxley studied zoology at Oxford, and in 1914 he started working at the newly established Rice Institute in Houston, Texas, as an assistant professor of biology. He returned to England after the start of the Great War, to serve his country as an army intelligence officer. For several years after the war he worked at Oxford, eventually becoming a well-known public figure. After the Second World War, he served first as secretary of the Preparatory Commission for the United Nations Educational, Scientific and Cultural Organization (UNESCO), and then as director-general of UNESCO.[34] Fascinated by animal behavior throughout his life, Huxley incorporated into his analysis of animal courtship both William Pycraft's interest in sexual stimulation and Edmund Selous's interest in sexual selection. Yet Huxley found female choice a problematic theory at best; the tendency of sexual selection to produce exaggerated plumage seemed to fly in the face of his fundamental belief in evolutionary progress through natural selection.[35]

In 1914, Huxley published "Courtship-Habits of the Great Crested Grebe," a long paper in which he avoided talking about the aesthetic preferences of the animals he studied.[36] He maintained that mating pairs of animals of different species formed a behavioral continuum: at one end was Pycraft's sexual excitement, "intuitive, unreasoned," and at the other, Selous's true choice, or "falling in love."[37] The instinctual model of mate choice described by Pycraft, Huxley continued, represented the "primitive condition" of mating behavior. In many animals, this primitive condition developed into choice-based mating systems. Increased male-male competition and female choice could lead, in highly polygamous and polyandrous species, to true sexual selection in the Darwinian sense. In monogamous species with a balanced

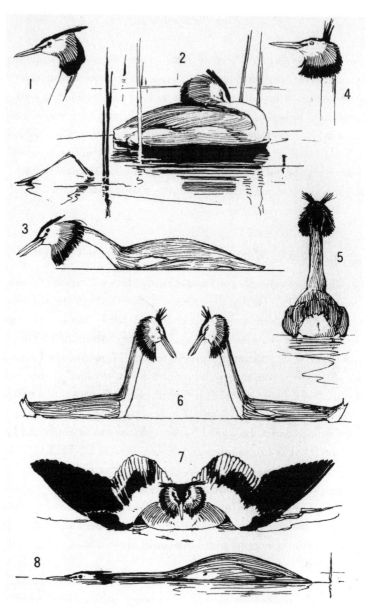

In the great crested grebe, according to Julian Huxley, males and females displayed to each other throughout the reproductive season, not just before mating. Both sexes participated in the so-called courtship dances. Huxley suggested that this form of courtship should be called "mutual selection." He denoted some of the various poses of the grebes during the day as follows: resting (1, 2), searching (3), shaking-attitude (4, 5, 6), cat-attitude (7), and the passive pairing attitude (8). From Julian S. Huxley, "The Courtship-Habits of the Great Crested Grebe (*Podiceps cristatus*); with an Addition to the Theory of Sexual Selection," *Proceedings of the Zoological Society* 35 (1914): 491–562, plate I.

ratio of males and females, however, the evolutionary development of choice-based courtship resulted in "mutual selection." In both cases, the excitatory function of display and courtship behavior had been discarded in favor of true mate choice.

Huxley thereby transformed Darwin's idea of a female's aesthetic comparison of males, through Pycraft's "emotional excitement," into a professional vocabulary and theoretical framework. In his discussion of animal courtship behavior, he replaced active comparisons, aphrodisiacs, and an aesthetic sense with "subconscious mental activities" and "inherited sexual passions."[38] Huxley still insisted, however, that animals of considerable mental complexity could and did make reasoned mate choices in a fashion that resembled courtship in humans.

Although Huxley's insistence on true choice-based behavior in animals was unusual in the 1910s, his construction of an evolutionary scale of behavioral complexity, from lower animals to humans, was not. In 1916, two Stanford zoologists, the aforementioned Vernon Kellogg and his colleague David Starr Jordan, co-published *Evolution and Animal Life*, in which they also argued that only higher animals might be expected to exhibit aesthetic choices.[39] They, too, concluded that the capacity to make aesthetic choices was insufficient for demonstrating female choice as an evolutionary mechanism. Additional proof was needed to show that mate choice was consistent both among individuals in the population and for an individual over its lifetime. Such evidence was lacking, they contended, especially in humans. Richard Swann Lull, a professor of paleontology at Yale, reiterated the argumentative points made by Kellogg and Jordan, and maintained that sexual selection did not explain anything not already covered by natural selection. He contended further that even if one could demonstrate an aesthetic sense in animals, "we have no assurance of any one standard of beauty, for individual taste varies greatly with men and probably also with animals." In effect, sexual selection was "the most doubtful factor of all those advanced by Darwin and is only held because nothing better has been offered in its place."[40]

About the same time, Julian Huxley, while a faculty member at the Rice Institute, presented a series of public lectures entitled "Biology and Man." In these lectures, he articulated a connection between animal and human courtship similar to Alfred Russel Wallace's view on human evolution. Huxley argued that proper female choice in humans would lead to continued evolutionary improvement in humans if women were treated as intellectual equals to men, given the vote, encouraged to participate in physical exercise, and allowed to think for themselves. Through the emancipation of women, mutual sexual selection would steadily improve human society.[41]

By the late 1930s, Huxley had revised his ideas on sexual selection.[42] In the years since publication of "Great Crested Grebe," he had become more convinced that sexual selection should be invoked as an explanation for sexual dimorphism only if natural selection could not explain the sex differences in question. He now described "threat (to rivals of the same sex) or sex recognition (by members of the opposite sex)" as the two most common functions of display—both of which he categorized as aspects of natural selection.[43] Huxley also proposed a host of other functions for behavioral display that obviated the need for sexual selection: warnings to predators, signals indicating sex recognition between juveniles and adults, and signals enabling sex recognition between adults at a distance.

In most cases, beautiful male display served as a signaling device and could be thought of exactly like a physical trait. "Display," Huxley contended, "is therefore analogous to a copulatory organ: it facilitates the union of the sexual cells, though it acts on the psychological rather than the purely physiological plane."[44] He accepted female mate choice as the underlying cause of a trait only when he could identify the trait as "beautiful rather than merely striking" or as "intricate or delicate." Huxley reasoned that for a display to be effective as a choice-inducing behavior, potential mates must have viewed the display at close range, where they could have perceived and evaluated differences among individuals. At far distances, such individual variation would be imperceptible. Therefore, the persistence of intricate patterns or behaviors in mating displays could not be attributed to distance recognition or threat posturing. the mating displays of most animals lacked such detail, however, and therefore could not be attributed to female choice.

As a result of his dissatisfaction with sexual selection, Huxley suggested replacing the outdated concept with two new ones.[45] The first, "epigamic selection," would denote "display characters common to both sexes" necessary for proper union of the gametes—such as the mutual displays of the great crested grebe. The second, "intrasexual selection," would incorporate all competition among members of the same sex for access to members of the opposite sex. In general, Huxley argued, beautiful male phenotypes helped overcome a female's natural sexual reticence and thus should be categorized as epigamic and therefore as natural selection: "in monogamous forms, all such epigamic characters can thus be regarded as owing their origin, at least in the main, to natural selection, but of a sex-limited nature."[46] Additionally, males often directed their displays at other males as a show of raw fitness, rather than directing their displays at females. Huxley included both epigamic and intrasexual selection under the umbrella of natural, not sexual, selection.

Despite Huxley's conviction that birds did not share humans' capacity for choice, he argued that they did share a preference for monogamy. Like Kellogg and Darwin before him, Huxley envisioned a possible role for female choice only in polygamous species with a male-skewed sex ratio, a mating system that he assumed was extremely rare. He noted that in monogamous birds, displays tended to be mutual and to occur after individuals had already paired for the season, thereby eliminating any theory that assumed a role for display in the process of mate *selection*. Huxley's normative views of human sexuality clearly played into this conviction that most animal species formed stable mating pairs through mutual monogamous stimulation. Often such pairings were restricted to a single mating season, with both male and female reproducing with different partners in the next season, stretching the strict equation of human marriage and animal courtship.[47] In his model of mating behavior, then, Huxley believed that animals used sexual display to excite a mate after pair formation, not only in the process of acquiring a mate. Moreover, it seemed to him that all animals would pair if they could, so there could be competition for females only when males were in "high excess."[48] Display in monogamous species, Huxley continued, should count as natural selection, as it occurred after pairing and was necessary for the propagation of the species. Only in polygamous species could intrasexual selection act to promote sex-specific traits used solely for the purpose of garnering the attention of the opposite sex. Even in polygamous species, however, most secondary sexual characters could be assigned a nonsexual function (warning to predators, stimulatory function, etc.) and so they, too, fell under the auspices of natural selection. In accepting monogamy as paradigmatic, Huxley found it difficult to envision a role for sexual selection in the evolution of male beauty.[49]

Huxley's 1938 papers were a thorough rejection of Darwin's evolutionary mechanism of female choice in animals and a drastic change from his earlier work on sexual selection in great crested grebes. His circumscription of sexual selection's efficacy to a small number of species reflected his larger caution about assuming mentality in animals. By arguing that beautiful male displays served to coax the female into sexual receptivity, Huxley short-circuited the problem of rationality and aesthetic choice in female mating behavior in animals. While he admitted that, in rare cases, female animals were probably capable of aesthetic choice, in practice he believed that they did not select their mates on the basis of this aesthetic appreciation but merely submitted to the most emotionally stimulating male.

Women, however, did make conscious choices about their potential partners. In 1931, Huxley co-published a biology textbook with Herbert George Wells, the famous historian and science fiction author, and George Philip Wells, Herbert's son and a biologist in his own right. In their textbook, these men drew a sharp line between human and animal courtship behavior: "The varied and individualized love-making of humanity differs fundamentally from the courtship of birds . . . The human lover may do a thousand things; the courting bird is an elegant determinate machine."[50] Although bird and human courtship behavior might be outwardly similar, they differed significantly in the underlying mental and emotional causes of behavior; animals do the same thing over and over, but humans "adapt their techniques to the occasion," because their behavior is controlled by different parts of the brain. The authors' differentiation between animalistic instinctual behavior and human adaptability and reflective choice provided a (neuro)biological explanation for why female choice and sexual selection could operate in human society only.

Even scientists interested in human sexual response found the topic of sexual selection and female choice compelling. Sexologists in the 1920s and 1930s intimated that just as in animals, women needed to be stimulated before each individual sex act. Noted sexologist Marie Stopes wrote in *Married Love* (1918) that "it should be realized that a man does not woo and win a woman once and for all when he marries her: *he must woo her before every separate act of coitus*; for each act corresponds to a marriage, as the beasts of the field and the fowls of the air know it."[51] Stopes suggested that just as animals would never mate without the rituals of courtship, neither should humans—especially as a man often wanted to have sex with his wife at various and multiple times, while a woman wanted to have sex with her husband only when naturally stimulated by her womanly cycle. "Wild animals are not so foolish as man; a wild animal does not unite with his female without the wooing characteristic of his race, whether by stirring her by a display of his strength in fighting another male, or by exhibiting his beautiful feathers or song." By interpreting human foreplay as animal courtship, Stopes's vision of the biological function of mating displays sympathized with Huxley's attempt to reconceive animal courtship as primarily stimulatory and not the basis of active female comparison. Yet, whereas Huxley argued that the human equivalent of animal courtship was monogamous human marriage, Stopes emphasized individual sex acts in humans. This connection between human and animal courtship was strengthened through the analogies used in a variety of contemporaneous marriage manuals—proper sexual etiquette in the marriage bed recapitulated animal courtship.[52] And, of course, vice versa.

Huxley's 1938 contention that, in animals, female choice of male beauty was nowhere near as common as Darwin originally postulated fit with the general trend of his zoological contemporaries to differentiate between human and animal mental attributes. When combined with experimental evidence from physiologists and geneticists that seemed to discount the possibility of female choice in animals, sexual selection no longer seemed a viable explanation of male beauty. Female choice relied on the very qualities that distinguished humans from animals: an aesthetic sensibility and intellectual reason. As Huxley's emphasis on display in animal courtship changed, so did the connections he drew between animal and human courtship.

Early in his career, when he thought mutual courtship display served as the basis of mate choice, Huxley emphasized the power of individual choice as a guide to the continued evolutionary progress of the human race. Later in his career, when he instead highlighted mating displays occurring after pair formation, Huxley postulated that in humans as in animals, mutual display was an important behavioral component of maintaining monogamous relationships by reinforcing the pair bond. Huxley remained extremely curious about the relationship between choice-based sexual behavior and evolution for the rest of his career. Although he never again published a research paper on the topic, he acted as mentor to other biologists interested in evolution and animal behavior.[53]

Female Choice, Marriage Selection, and Evolutionary Theory

Starting in the 1920s, three men—Ronald Aylmer Fisher, J. B. S. Haldane, and Sewall Wright—began to integrate genetics with natural selection, using mathematics to describe the evolution of a population.[54] Only one of these mathematically inclined evolutionary theorists paid significant attention to female choice: Ronald Fisher. After Fisher was rejected for fighting duty in the First World War, because of his terrible eyesight, he decided he could best serve his country by becoming a farmer. In 1917, Fisher and his new bride, along with her older sister (a long-time friend of Fisher), moved to a former gamekeeper's cottage in Bradford, England. They were soon to be joined by several pigs, at least one calf, and children. Fisher sought to exemplify a eugenic life through subsistence farming, and it was only after the war that financial constraints, along with the continued exhortations of Charles Darwin's fourth son and eighth child, Leonard Darwin, to give up farming, led Fisher to accept a post as a statistician at Rothamsted Experimental Station. The early years of his professional career exemplify the passions that would last throughout his life: eugenics, theoretical population biology, and statistical analyses.[55]

Fisher's interest in female choice was part and parcel of his more general explication of sexual selection and the evolution of humans in *The Genetical Theory of Natural Selection*, published in 1930.[56] In *Genetical Theory*, Fisher used sexual selection to argue that female choice could explain the evolution of apparently nonadaptive traits in humans, such as self-sacrifice (heroism) and beauty. These two traits presented a problem for any evolutionary theory based solely on natural selection, he believed, as neither could be explained as enhancing the survival of individuals. Fisher's concern with the physical fitness and vigor of the next generation also drove his interest in sexual selection. Far from trying to remove the taint of "vigor" from theories of natural selection, Fisher intended his theories of natural selection and sexual selection to demonstrate the importance of eugenic concerns to the physical and intellectual prowess of the next generation. Fisher's commitment to female choice and positive eugenics was explicit in his early publications in the *Eugenics Review* (for which he also acted as book review editor for many years), especially "Some Hopes of a Eugenicist" (1914), "The Evolution of Sex Preference" (1915), and "Positive Eugenics" (1917).[57] His later elaboration of sexual selection described the process of evolution as based on differential reproduction, occurring within a single population, and resulting in either progressive development or degradation of the population.

Fisher's formulation of sexual selection and female choice in *Genetical Theory* emerged from both his genetic and evolutionary interests in human progress.[58] His interest in eugenics began early in his education, and as an undergraduate he was elected the first chairman of the Undergraduate Committee for the Cambridge Eugenics Society. Fisher, like Darwin, believed that females compared males and chose the one they preferred. The belief that female choice was a comparative, choice-based evolutionary mechanism lay at the heart of Fisher's conviction that sexual selection was more important for the improvement of humanity's biological future than it was for animal evolution.[59]

In his short 1914 paper, "Some Hopes of a Eugenicist," Fisher set forth Darwinian evolution as vital to the future of humanity. "From the moment that we grasp, firmly and completely, Darwin's theory of evolution, we begin to realise that we have obtained not merely a description of the past, or an explanation of the present, but a veritable key of the future." Selection influenced the physical body of man as well as his mental and emotional capabilities. "All the refinements of beauty, all the delicacy of our sense of beauty, our moral instincts of obedience and compassion, pity or indignation, our moments of religious awe, or mystical penetration," were the result of evolution in action.[60]

Fisher identified at least two facets of the eugenic problem: the quality and the quantity of offspring in the next generation.[61] In choosing with whom to reproduce, humans' marriage choice determined the quality of children in the next generation. The number of babies born to families of high and low quality determined the proportion of high- and low-quality adults in the next generation. In 1915, Fisher addressed the first of these aspects, quality of offspring, by linking the past and future evolution of humans to mate choice in "The Evolution of Sexual Preference." In this article, Fisher contrasted the effects of Darwin's theory of natural selection with the effects of sexual selection. Natural selection, he argued, could explain the evolution of the organization and physical structures of animal and human bodies. The evolution of human ethics, aesthetics, beauty, and morality he reserved as the special jurisdiction of sexual selection. "In the sexual selection of mankind, therefore, beauty and character provide standards of universal currency. Everywhere we see the power of beauty of form, colour, voice, expression and grace of movement; everywhere also the charms and merits of character dominate our judgments." Further, human capacity to perceive all the "elusive traits of human excellence" was at its most acute under the influence of burgeoning love.[62] Only two years later, in "Positive Eugenics," Fisher addressed the second issue—quantity of offspring. He argued adamantly that upstanding professional families needed to produce more children. "To increase the birth-rate in the professional classes and among the highly skilled artisans would be to solve the great eugenic problem of the present generation and to lay a broad foundation for every kind of social advance."[63] Fisher's vision of eugenic evolution in human society, then, included both the quality and quantity of the offspring who would form the next generation.

Much like Wallace, Fisher expressed skepticism that evidence of sexual selection in nonhuman animals would ever be found. He insisted that although the process of sexual selection in animals and humans was very similar, "among mankind the conditions are more favourable for the higher development of sexual selection . . . [because] the choice of a mate is of more importance among mankind than among most other animals."[64] Not all humans were equally good at judging mate quality, however; individuals of high intellectual development could best appreciate fine differences in the quality of potential mates. Fisher's linkage of aesthetic sensibility and intelligence gives us a sense of how biologists used sexual selection to both explore the animalistic basis of human sexuality and simultaneously reserve for humans their highest expression.

These factors of mate quality, offspring quantity, and the importance of individual choice in directing human evolution came together in Fisher's description of sexual selection in *Genetical Theory*. The book had twelve chapters. The first seven explored evolutionary theory in general; the last five dealt more specifically with humankind. In the preface, Fisher indicated that "the deductions respecting Man are strictly inseparable from the more general chapters."[65] The first chapters contained the mathematical and theoretical foundations of Fisher's later arguments about social economic reform. He most clearly explored the connection between animals and humans (or, more generously, between the general theory and its specific application) in chapter 6, "Sexual Reproduction and Sexual Selection," and chapter 11, "Social Selection of Fertility."

From a historical perspective, Fisher at first seems to be a bit of a renegade within the biological community because he argued, against the wisdom of contemporary zoologists such as Kellogg, that sexual selection could work in animals. Yet, like his contemporaries, Fisher emphasized the power of sexual selection and female choice to enact evolutionary change in human societies rather than in animals. In his chapter on sexual selection in *Genetical Theory*, Fisher outlined a process in which female preference for a trait (say, long tails in birds) could cause such a trait to spread in a population, even if it had no survival value for the males or, even worse, hurt the males' chance of survival. Fisher called this process "runaway sexual selection." Today, biologists interested in animal behavior and evolutionary theory look back on Fisher's work as seminal to the development of their field, but his formulation of runaway sexual selection failed to attract much attention among biologists until the 1960s.

For Fisher, the power of sexual selection rested in its ability to act in opposition to the selective effects of natural selection. For female preferences to become ingrained in a population, Fisher posited, initially there must be some fitness advantage inherent to females' choices.[66] He suggested there were only two necessary conditions for directional evolution through sexual selection: first, that a sexual preference must exist in at least one sex, and second, that the preference must confer a reproductive advantage. Once heritable mate preferences were established in a population, they could begin to act as a mechanism for evolutionary change in their own right. Given these conditions and enough time, sexual selection could even help generate traits that were detrimental to individuals survival—hence, "runaway" sexual selection.

According to Fisher, runaway selection occurred when female preference for a male trait and the male trait itself began to evolve in lockstep. Fisher used the example of plumage development in birds to illustrate his point. As some males' plumage became more beautiful and extravagant (think of a peacock's tail), the effect of the average male's plumage on females declined—the fancy males would produce many more offspring than the plainer males. The next generation would disproportionately exhibit longer and brighter tails. If females continued to choose to mate with males with the longest and brightest tail, their preference would become more intense as the male traits became more extravagant. Even if extravagant males were less likely to escape predators (natural selection), further development of male tails would proceed until the reproductive advantage was completely negated by the disadvantage incurred by natural selection. Thus, through female choice, populations would evolve in ways that carried no survival advantage (or even decreased individuals' chances of survival) but increased the ability of females to discern and appreciate beauty.[67]

Fisher used his illustration of runaway sexual selection in birds to elucidate the evolution of apparently nonadaptive traits in humans, such as heroism. Heroism, for Fisher, described the tendency of people (especially men) to sacrifice their lives in battle for the good of the group (tribe) to which they belonged. Fisher found it difficult to explain such altruistic tendencies solely in terms of natural selection, which he thought should act to cull such phenomena from a population very quickly. He compared the phenomenon of heroism in humans with the presence of bitter-tasting chemicals in some insects—the evolutionary benefit of both traits was conferred on the group to which such individuals belonged.[68] Just as the bitter-tasting butterfly earned no personal fitness advantage (no increase in number of offspring) when eaten by a predator that subsequently learned to avoid eating similar-looking butterflies, the warrior's sacrifice of his life for the safety of the group also connoted no personal advantage.[69] Fisher argued that both of these traits were maintained evolutionarily by the reproductive advantage gained by the *relatives* of the sacrificed individual—a form of what biologists now call "kin selection."[70]

In one major aspect, however, the evolution of bad-tasting insects and heroic humans differed for Fisher—sexual selection. The effects of female choice and sexual selection complicated explanations of human heroism, because women preferred to mate with heroic men. If the hero did not die but returned home safely, his exploits would surely make him the focus of many women's attentions. Fisher hoped that proper female choice in this context would correct the dysgenic effects of war. When his son died in combat in 1943, Fisher searched in vain for evidence of an unreported grandchild, hoping that his son's genetic heritage would not be lost.[71] Part of Fisher's concern over the loss of his son was a specific instantiation of a much more widespread concern about the impact of the Great War on the United Kingdom's population: war eliminated countless thousands of the bravest young men from the breeding population. Female choice for heroes ensured the eugenic quality of the next generation, but only if those heroes produced offspring.

Sexual selection could also enhance the effects of natural selection, and it served both as a viable method of eugenic improvement and as a cautionary tale about the inevitable effects of the fertility trends of modern society.[72] As long as the preeminent members of society were more fertile than less desirable members, and women continued to choose their husbands well, then society would become more intelligent and more beautiful. If, however, the relationship between social and sexual selection for heroes and fertility were reversed, sexual selection would act to decrease the number of heroes in the population, and the numbers of less valuable members of society would increase. "Sexual selection must be judged to intensify the speed of whichever process, constructive or degenerative, is in action. The intensity of this influence must, however, be much diminished in the later stages of civilized societies, with the decay of the appreciation of personal differences."[73] Fisher argued that as the educated elite gave birth to fewer children than the masses, the connection between fertility and eugenic quality was becoming decoupled. He feared that in the future, women would lose the ability to choose well, and ill-advised female choice would speed social destruction.

Outspoken eugenicist William J. Robinson, who provided an extreme example of the evolutionary ladder of aesthetic appreciation and intelligence, expressed similar concerns about the lack of eugenic mate choice in civilized society. In his immensely successful *Woman: Her Sex and Love Life*, Robinson distinguished between the sexual drive to mate, which intercourse with any member of the opposite sex can satisfy, and feelings of true love for one individual above all others.[74] "Real love, true love, is a new feeling, a comparatively modern feeling, absent in the lower races and reaching its highest development only in people of high civilization, culture, and education."[75] Robinson contended that only civilized people could exhibit true love and rational female choice; members of the lower races, just like animals, failed to differentiate their sexual interests. Similarly, in a 1937 article on teaching women proper mate choice, prolific American social hygienist Paul Popenoe admonished educated women to follow their evolutionary past in attracting a mate, by acting "seductive and alluring rather than aggressive" so as to elicit male courtship rituals. To behave otherwise, Popenoe suggested, was as dangerous to marital bliss as were insufficiently aggressive husbands.[76] Through rational analysis and change of behavior, Popenoe contended, women could become more attractive mates and better wives.

Whereas books by Robinson and Popenoe sold thousands of copies, Fisher's *Genetical Theory* attracted far less popular and scientific attention. After publication of the book, Fisher was dismayed that few reviewers commented on the application of his general theory to the problem of human evolution. In a letter to geneticist J. B. S. Haldane, Fisher wrote of his work that "the main practical point is to combat the idea that racial decay, or the differential birth-rate, or any other social phenomenon which we judge undesirable, is to be accepted fatalistically as the 'Will of Allah,' rather than tackled scientifically like rabies." Fisher had hoped that the practical application of his theories to social policy would attract more attention. In a similar letter to geneticist Hermann J. Muller, Fisher lamented that "I feel on rather strong ground about Man, apart from the fact that one cannot make any innocent statement about that confounded animal without people thinking you are attacking their political or religious opinions." He continued by pointing out that his first chapters on humans, "Chapters VIII to XI [,] are an attempt to build up one coherent argument analogous to sexual selection, or mimicry theory (I mention this because the best critic whose opinion I have yet heard, has missed just this point)."[77]

In fact, the entire *Genetical Theory* failed to achieve much critical acclaim among

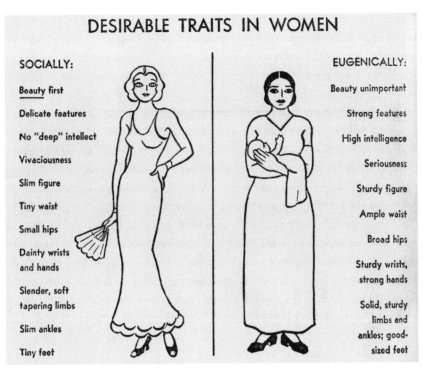

Ronald Fisher, William Robinson, and Paul Popenoe all worried that mate choice gone wrong would be devastating for human evolution. This image from a popular text on genetics emphasizes that the result of choice for beauty alone yielded socially well-adapted but evolutionarily useless women. From Amran Scheinfeld, *You and Heredity* (New York: Frederick A. Stokes Company, 1939), 571.

biologists until the second edition appeared in 1958. One explanation is that the mathematics associated with the population dynamics of the early chapters was quite difficult; as a result, perhaps very few readers made it to the later chapters of the book.[78] Another explanation could be that readers shied away from the book because of its explicit political agenda. In either case, biologists' failure at the time to cite Fisher's runaway selection model of female choice was probably not due to their lack of interest in female choice or behavior as an evolutionary mechanism in *animals*, but simply a reaction to the book as a whole.

The lukewarm reception of *Genetical Theory* did not signal the end of Fisher's interest and research in human evolution. To be sure, he became frustrated with the decreasing importance of research in the Eugenics Education Society, and he ended his association with the group.[79] His continued interest in the questions of human evolution and genetics is evidenced by his serological work in the 1930s and 1940s.

Through an analysis of blood groups (ABO) and the rhesus (Rh) factor, Fisher hoped to elucidate the relationship between racial blood types and susceptibility to particular diseases.[80] Despite his continued interest in racial evolution and genetics, Fisher never returned to the issue of female choice or sexual selection in his published research.

Fisher's subsequent lack of publications on female choice also seems to have stemmed from his conviction that he had solved the problem of mate choice in humans. His reaction to a 1948 paper, "Intra-sexual Selection in *Drosophila*," speaks volumes about his later attitude to research on female choice and sexual selection in animals.[81] Fisher knew of the author, Angus John Bateman, because Bateman worked at the John Innes Horticultural Institute, under the direction of cytologist Cyril Darlington.[82] Darlington and Fisher were long-time friends, cofounding (in 1947) and coediting the journal *Heredity*. Bateman's paper, published in the second volume of *Heredity*, proposed that proper mate choice was more important to females than males, because females invested more resources in each young and mated far less frequently than did males. If a female made one poor mate choice, Bateman theorized, it could affect her entire reproductive contribution to the next generation; if a male made one poor mate choice, it affected his reproductive success far less, because his choice was only one among many in his lifetime. On reading the manuscript, Fisher made an off-hand remark to Darlington, who passed the comment along to Bateman. In response, Bateman wrote to Fisher protesting that he had not, in fact, plagiarized Darwin and insisting that his work represented a new framework within which to analyze the importance of female choice and sexual selection—namely, reproductive investment.[83] Fisher immediately attempted to clear up the misunderstanding.

> I am sure Darlington misunderstood me if he suggested that I thought you guilty of plagiarism. Of course that would be nonsense. I certainly had supposed it was obvious, and had been taken as obvious by writers on the subject, that the reproductive capacity of females being more limited than that of males, it was inevitable that they rather than the other sex should exercise selection and that the males should be more conspicuously modified by sexual selection . . . However, I do not think all this matters; at least your paper makes sure that future writers will be able to find the general principle clearly stated.[84]

How reassuring, and yet how dismissive! For Fisher, the crux of female choice and sexual selection was *human* evolution. Working through the dynamics of female choice as it related to *animals* was trivial and missed the point. Through the lens of human mate choice and eugenics, it seemed obvious to Fisher that the differential investment of the sexes in reproduction and childcare meant that women should consider only a subset of men to be worthy mates, while men could afford to be significantly more profligate in their sexual tastes.[85] For Bateman, looking only at animals, the connection was new and subtle.

In Fisher's view, evolution was a directional process, potentially progressive or degenerative, taking humans closer to or farther from animality, and acting on a single population. Female choice in marriage partners led to differential reproduction, in which some males more than others contributed to the genetic composition of the next generation. Such differential reproduction acted as an evolutionary accelerant, either improving the eugenic worth of a single breeding population or hastening the population's demise through genetic deterioration. Fisher appears to be a biological renegade in promoting female choice and sexual selection as valid biological concepts only if we ignore the human context of his evolutionary theorizing.

In the first decades of the twentieth century, scientists interested in evolutionary theory believed that by studying reproductive behavior in animals, they might learn about the biological basis of our own sexual desires. Their reactions to female choice as a potential mechanism for evolutionary change illuminate the historical divergence of two models of animal courtship, one that emphasized the stimulatory and bonding functions of sex and foreplay in human marriage, the other that analogized animal courtship and the process of choosing a spouse. These two models of animal-human courtship contained different assumptions about the function of extravagant displays in animals: Was the observer emotionally stimulated to react? Or did the display provide an empirical basis for the observer to judge the evolutionary "worth" of the displayer?

Charles Darwin originally proposed a mentalistic model for female choice, in which females actively compared the males around them and chose the one they found the most attractive or beautiful. This choice-based model caused considerable difficulty for naturalists who wished to draw a sharp line distinguishing human cognition from animal emotion. Decades later, Thomas Hunt Morgan and Alfred Sturtevant dismissed fruit flies as incapable of such reasoned choices. They suggested instead that differential male vigor and ability to stimulate the female into a mating frame of mind was behind apparent female choice. Yet not all geneticists, or population geneticists, adhered to Morgan and Sturtevant's stimulation-based model, just as not all naturalists advocated Darwin's choice-based model.

Ronald Fisher remained focused on the human dimension of his evolutionary theorizing throughout his career. In his research, Fisher never worked directly with experimental organisms and was criticized by some biologists for his lack of zoological training. In 1931, naturalist Ernest McBride was reported as exclaiming that "the opinions of Dr. Fisher are of no value since he is not a trained Biologist, and a mathematician's ideas are of no importance on the problem of population!"[86] Yet most biologists did take Fisher's mathematical explanations of populations seriously, even as they overlooked the intended target, human evolution. Because of the human context for Fisher's mathematical theories, he found no difficulty in subscribing to a choice-based model of sexual behavior. At the same time, he despaired of ever finding demonstrable proof of sexual selection in animals, precisely because he also adhered, like Huxley, to the belief that most animals did not have the intellectual capacity to choose their mates.

Julian Huxley, a naturalist by training, argued that most extravagant plumage in male birds served functions that could be ascribed to natural selection, such as delimiting a territory, species recognition, and sex recognition. He maintained that choice-based behaviors accounted for male beauty in a very few instances—only if the breeding population was polygamous *and* contained many more males than females. In such circumstances, even if all the breeding females found mates, some males would not reproduce. Thus female choice of mates could have a large effect on the next generation. In populations with equal numbers of males and females, he reasoned, all reproductively mature individuals would end up breeding, and female choice of mates, in the end, would have no effect on the composition of the next generation. Huxley also insisted that most species were monogamous and contained approximately equal numbers of males and females. Therefore very little beauty in the animal kingdom could be reasonably ascribed to choice-based behavior.

Zoologists' reactions to female choice and sexual selection in the first third of the twentieth century also depended fundamentally on their understanding of how choice operated in a population. Both Huxley and Fisher believed that consistent choices in mates over time, resulting in differential reproduction, could theoretically drive the evolution of physical and behavioral differences between males and females of the same population. As a part of this belief, they also agreed that evolutionary change acted within a *single population* to produce heritable changes in that population. Neither sought to understand how behavior might affect the process of speciation. When Fisher couched natural selection in mathematical, statistical terms, he instantiated a common vision of evolutionary change as based on differential reproduction, within a single population, involving directional change over time. The framing of evolution as a process affecting a single population helped facilitate analogies between animal species and humans.

Female choice and sexual selection also served as popular modes of evolutionary change in human populations in part because differential survival of individuals in a population (natural selection) did not seem to be as much of an issue for "civilized" society as it was for more "primitive" peoples, or for animals. If anything, natural selection in modern society had been reversed by technological "total war," as the most "fit" individuals died in greater proportions than less desirable individuals. While proper female choice could ensure the quality of offspring, eugenic families also needed to produce more babies than dysgenic families, so that the next generation would derive disproportionately from the more favorable genetic stocks. When applied to humans, female choice, sexual selection, and better babies programs seemed to promise a rational, controlled means of changing society.[87]

The influence of social hygiene, positive eugenics, and a belief in human progress preceded and directly influenced early twentieth-century mathematical formulations of evolutionary theory and animal mating behavior. Certainly, by the late 1930s, British and American biologists attacked the biological basis of negative eugenic measures (those aimed at reducing the rate of reproduction of the lower races and other social undesirables through sterilization). These attacks came from across the political spectrum—from Fisher, the "antiracist conservative," to "social radicals" such as Havelock Ellis and George Bernard Shaw (all three of whom, it is worth noting, based some of their social arguments in sexual selection theory).[88] In the place of negative eugenics, these proponents of "reform," or positive eugenics, advocated increasing the reproductive rate of upstanding citizens. Yet, new theories of population genetics did not banish the specter of biological theory applied to human populations—far from it. At least for Fisher, positive eugenic theory served as a model for how to fruitfully combine the concept of reproduction (quality and quantity of offspring) with the question of evolution (genetic change in a population over time).

In the early twentieth century, then, biologists' common vision of evolution as a directional process affecting a single population provided an easy analogy between human and animal evolutionary processes. By the 1940s, that analogy had already begun to dissolve as biologists adopted a new framework of evolutionary change based on nondirectional speciation.

CHAPTER 3

Branching Out, Scaling Up

American Experiments on Behavior

The strange ways of courtship and mating among fishes, frogs, lizards, snakes, birds, to say nothing of rats, guinea-pigs and monkeys, used to be recorded separately by ichthyologists, herpetologists, ornithologists, mammalogists, primatologists and comparative psychologists all writing in as many different technical journals, which in turn were rarely seen by anyone outside their respective specialties. More than any other man [Gladwyn Kingsley] Noble was rapidly integrating these fragments into a continuous and understandable picture . . . His ideal [was] the discovery of the basic principles that have governed the evolution and behavior of vertebrate animals from fish to man.
—William K. Gregory, Address at Noble's memorial, 1940

The study of animal behavior in the United States expanded considerably between the two world wars, in terms of the number of biologists interested in the subject and the scope of their research.[1] Biologists working on animal behavior employed techniques from both naturalist and experimentalist traditions. For example, Warder Clyde Allee, at the University of Chicago, incorporated animal behavior into an ecological context. Allee stressed the importance of an organism's interactions with the community in which it lived and with the surrounding environmental conditions in producing its behavior.[2] Approaching behavior on a different tack, psychologist William C. Young, one of the founders of behavioral endocrinology, studied the role of sex hormones in producing mating behavior. Both of these trends in behavioral research of the interwar period built on strong disciplinary traditions in the study of behavior established at the turn of the century. Additionally, by the mid-1930s, comparative psychology began to attract more students. Comparative psychologists were interested primarily in the ability of animals to learn, although some also explored the role of behavior in the natural lives of organisms and the evolution of behavior more generally.[3]

Despite their methodological diversity, these biologists were united in their belief that the study of animal behavior should be professionalized and rid of its amateurish, anthropomorphic roots. Each group brought valuable contributions to the table: experimentalists found that the controlled environment of the laboratory provided an ideal location for modifying and observing behavior in developing organisms, while naturalists used observation and modification of an organism's natural habitat to establish the normal behavioral characteristics of the species. Without the behavioral data gathered in the natural environment, it was impossible to know whether or not behaviors observed in laboratory spaces were simply artifacts of the artificial, laboratory conditions. As a result, the activities of naturalists and experimentalists overlapped and complemented each other within the professional study of animal behavior.[4]

Biologists' understanding of evolutionary theory changed dramatically in this period. In the early twentieth century, practicing zoologists had adopted an evolutionary model in which many different fauna and flora adapted simultaneously and independently to their surroundings. The process of evolution, then, drove a single species in a particular direction (for example, toward more refined adaptation to its environment or toward greater sociability). By the 1950s, American studies of animal behavior looked remarkably different. Zoologists portrayed evolution as fundamentally concerned with the process of speciation, in which subpopulations of a single species broke off and assumed a distinct evolutionary future of their own. Whereas in a linear model of evolution it was easy for biologists and nonscientists alike to see human cultures as the pinnacle of social evolution and animal behavior as a primitive form of human behavior, in a branching model, the comparison between animals and humans was much less clear.

Given these dramatic changes in the communities of scientists interested in animal behavior and evolutionary theory during the first half of the twentieth century, it is hardly surprising that biologists also altered their conceptions of female mate choice in animals. Biologists in the 1930s conceived of female choice as a cognitive comparison, one female from a particular species choosing from among multiple male suitors of the same species, and they supported or dismissed the evolutionary mechanism accordingly. By the 1950s, biologists instead couched female choice as the process by which females in a population determined the species identity of their potential suitors and were stimulated to mate by any male exhibiting the appropriate species-specific courtship behaviors. This transition, from female choice as individual comparison to female choice as response to the courtship characteristics of a group of individuals, is illustrated vividly by the line of curators in charge of animal behavior research at the American Museum of Natural History (AMNH) in New York.

Gladwyn Kingsley Noble, curator of herpetology at the AMNH from 1923 to 1940, embodied the fluid boundary between experimental and naturalist traditions characteristic of animal behavior research in the 1930s. At the AMNH, Noble founded the Laboratory of Experimental Biology (LEB) in 1928. In his laboratory spaces, he sought to map the evolution of social behavior across all vertebrate animals and to uncover the hormonal changes regulating the unique courtship behaviors of each taxonomic group he studied.[5] To do so, Noble observed and gathered his research specimens in the field, using the field data to construct naturalistic enclosures for his experimental subjects in the LEB. In 1939, Noble published an article entitled "The Experimental Animal from the Naturalist's Point of View," in which he extolled the virtues of laboratory research for answering questions of concern to naturalists like himself.[6]

Noble had grown up in Yonkers, leaving the New York area for undergraduate study at Harvard University. By the time he graduated, in 1917, he was working with up-and-coming herpetologist Thomas Barbour at the Museum of Comparative Zoology and had already conducted field research in Guadeloupe, Newfoundland, and Peru. He intended to stay at that museum and obtain at least a master's degree, but fears of being sent to fight in Europe and failing family finances led him to seek employment closer to home—at the American Museum of Natural History. Noble's "abrasive personality and tremendous ego" did not make him everyone's favorite curator at the museum.[7] With a few leaves of absence (military service, after all, and more training under Barbour at Harvard), Noble eventually earned his Ph.D. at nearby Columbia University and worked at the AMNH until his sudden death in 1940.[8]

After Noble's death, the LEB's experimental research program began to change direction as subsequent curators strove to fit their experimental research on behavior into the overall mission of the AMNH. Frank Ambrose Beach, curator of the laboratory from 1940 to 1946, continued its comparative approach to behavioral research and changed the name of the laboratory to the Department of Animal Behavior. This change codified and made explicit what had become the sole research agenda of the LEB under Noble's tenure. When Beach left, Lester Aronson took his place as curator of the Department of Animal Behavior. Like Noble, Aronson initially worked to show that experimental research on animal behavior was valuable to the other naturalists who worked at the museum. Aronson's research into evolution and reproductive behavior, however, differed considerably from Noble's in its theoretical and methodological framework. Whereas Noble emphasized the *pattern* of behavioral development over evolutionary time, Aronson explored the role of mating behavior within the *process* of speciation.

This shift from pattern to process in the explanatory goals and experimental design of animal behavior research at the AMNH reflected the increasing influence of experimental population genetics in shaping the research practices characteristic of mid-twentieth-century evolutionary biology. The most important theoretical development affecting biologists' ideas about evolution and sexual behavior was the new interest among population geneticists working in the United States on reproductive isolation. During the late 1930s and early 1940s, works by geneticist Theodosius Dobzhansky, zoologist Ernst Mayr, and paleontologist George Gaylord Simpson brought natural selection to the fore as the most significant causal mechanism governing speciation and macroevolutionary change in animals.[9] American biologists interested in behavior and evolution changed the focus of their research from the evolution of the behavior of individuals to the frequency of behaviors in a population.

These two models contained different assumptions about the relationship of animal and human behavior. In Noble's progressive evolutionary scheme, the behaviors of animals served as models of prehuman social interactions. In the work of later population geneticists and in Aronson's branching evolutionary model, the behaviors of extant animals and humans were equivalent in evolutionary time, each adapted to its unique environment. Within such a model, the human equivalent of animal courtship (or the animal equivalent of human courtship) became more problematic, because zoologists did not typically conceive of humans as being in the process of speciating.

Thus, in the United States from the 1930s through the 1950s, as biologists' understanding of the evolutionary process changed, so too did their experimental designs, methodologies, and research questions. Their research focus shifted from uncovering the evolutionary pattern of behavioral development to understanding how behavior might affect the process of evolutionary change.

Founding the Laboratory of Experimental Biology

On the west side of Central Park in New York City, Gladwyn Kingsley Noble created what he hailed as a unique research environment. Construction of the Laboratory of Experimental Biology began at the American Museum of Natural History in 1928, and the laboratory formally opened in May 1933. In a report for that year's meeting of the trustees, Noble described the opening as "the most significant single event of the year . . . this gives the American Museum the finest experimental laboratories in any museum of the world."[10] Although collecting and displaying live animals for the purposes of entertaining and educating the public had been relatively widespread in museums since the nineteenth century, the use of live animals as research organisms at a museum was far less common.[11] Experimental research of this sort was more likely to be conducted at a university or zoo.

Modeled after Hans Przibram's Biologische Versuchsanstalt in Vienna, the finished product included rooms for aquaria, greenhouses, animal breeding, balance apparatus, darkrooms, cold rooms, laboratories for histology and physiology, and facilities for sterilizing equipment.[12] Six investigators could work in harmony. Construction of the laboratory was largely funded by money given to the museum by the City of New York and private donors for the construction of the African Wing— the laboratory occupied the uppermost floor of the new wing.[13] The total bill was approximately $82,000, a substantial investment in 1933, but small when compared with the museum's expenditure on field expeditions during the same period. From 1928 to 1933, the years of the laboratory's construction, the AMNH spent more than $263,000 on the Central Asiatic Exploration and Research Fund.[14] Adjusted for inflation, the construction of Noble's laboratory would cost approximately $1 million in 2008 U.S. dollars.

Noble's research agenda provides a good idea of how he intended his laboratory to be integrated into the museum as a whole. As the laboratory opened its doors, he outlined fourteen lines of research integrating natural history and experimental zoology.[15] He hoped that in collaboration with the laboratory, other curators could solve questions they could not answer using data gathered solely in nature. For example, he believed that studying the hormonal basis of tooth formation would prove of great interest to vertebrate paleontologists and zoologists. Dental morphology often forms the basis of classification in vertebrates because tooth enamel is more durable than bone; even when the rest of the animal has long ago turned to dust or soil, the teeth often remain. Noble also argued that species identification could be accurate only if tested with controlled breeding experiments establishing the genetic constitution of putative new species or subspecies collected in nature. He further hoped that by injecting specimens of rare or cryptic species with concentrated extract from their pituitary glands, he could induce them to breed out of season, thus allowing naturalists to study their life history for the first time. Noble wanted the LEB to act as a central research space serving the needs of other departments in the museum. The AMNH trustees supported the construction of the laboratory with exactly this hope—that it would engender cooperation among the traditionally fractious departments.[16]

As the economic prospects of the country continued to decline, it became clear to Noble that he would have to streamline his research to accommodate budgetary restrictions. In 1933, the AMNH trustees decided to halt all institutional expenditures on field expeditions and concentrate their resources on museum-based research. (Privately financed field expeditions still occurred if the necessary funds had been donated to the museum with the stipulation that they be used only for a specific expedition.) The ability of the museum to continue functioning during the Great Depression was in large part due to the Works Progress Administration (WPA) and the museum's status as a semi-public institution. For example, in 1936 the WPA granted the museum sufficient funds to employ almost 250 WPA workers. Of these, about 60 were assigned to Noble for work in translation of research papers, exhibit construction, and laboratory research. Although these funds were sufficient to keep the museum's exhibits open and the laboratory functioning, they were not sufficient for purchasing consumable laboratory supplies. Noble also applied to the Josiah Macy Foundation and to the National Research Council's Committee for Research on Problems of Sex (NRC-CRPS) for money to acquire research organisms, chemicals, cages, and food. Through this combination of outside grant money and WPA workers, research at the laboratory continued and Noble's team turned out papers.[17]

Noble also chose to limit himself to what he considered the most "urgent" of his research plans—the evolutionary history of the psychological basis of behavior.[18] Noble's work on sexual selection formed one of the central aspects of his research plan. Referring to his research on the significance of bright colors in fishes, lizards, and birds, he wrote: "we have investigated this question both in the laboratory and the field with the result that we have been able to advance new views as to the significance of these adornments. Further observations are desirable on other species before these views will receive general acceptance."[19] He hoped that later, when financial conditions in the country improved, he would be able to expand the scope of his research to include some other questions of the fourteen he had originally intended to answer.[20] The laboratory, however, never stimulated the interdepartmental cooperative spirit for which Noble and the trustees had hoped.

Although Noble was originally hired into the Department of Herpetology, in his laboratory he considered all vertebrates part of his scientific jurisdiction, and he published papers on fishes and birds in addition to the typical herpetological gamut of lizards, chameleons, turtles, snakes, and tadpoles. In 1937, he hired Frank A. Beach to be assistant curator of the LEB and further extended the reach of the department into mammalian physiology.[21] Early in his career, Noble argued that "it is evolution which makes zoology a unified science and the student of zoology at the outset of his career should be given the opportunity of glimpsing the whole edifice of animal life before being called upon to analyze the functions of its various parts."[22] This evolutionary framework formed a central precept of his research on reproductive behavior.

Noble envisioned his interests in mating behavior and systematics as two aspects of the same enterprise. He wanted to know how mating behavior developed in evolutionary time. Using taxonomy as a proxy for time, he hoped to trace the development of sexual behavior from the lowest to the highest vertebrates—namely, humans. Once he had established what should count as "natural" behavior in a species from each major vertebrate group, he hoped to use laboratory techniques to elucidate the physiological mechanisms underlying the expression of those behaviors. Noble followed the same research plan with each species he investigated. The final product would have been a map of the evolution of hormonal and neural control of natural reproductive behavior in vertebrates, illustrating the increasing complexity of sexual behavior throughout Vertebrata.[23]

When Noble died suddenly at the age of forty-six, of a vile streptococcal infection known as Ludwig's quinsy, he had not yet completed his work. At the memorial, William Gregory commended Noble for his broadly comparative investigations of "the strange ways of courtship and mating among fishes, frogs, lizards, snakes, birds, to say nothing of rats, guinea-pigs and monkeys." Noble, Gregory continued, had been "rapidly integrating these fragments into a continuous and understandable picture . . . [to discover] the basic principles that have governed the evolution and behavior of vertebrate animals from fish to man."[24] The phrase "from fish to man" was a favorite of Gregory's, and in retrospect it seems likely that he used his memorial address as an opportunity to advance his own vision of how to unify research at the AMNH around evolutionary questions.[25] Yet the two men's analytical systems were similar—Gregory's research centered on the physical, morphological evolution of vertebrates, and Noble's research traced the sexual and hormonal evolution of behavior in vertebrates. Noble had advanced only as far as the first or second step in his intended research program with some of the species he planned to work on, but his hopes of integrating evolution and behavior into a single analytical framework reflected his own attempt to arrive at a synthetic biological knowledge.[26]

After Noble's death, his widow, Ruth Crosby Noble, gathered up what she could find of his research notes and transformed them into a popular book, *The Nature of the Beast*, so that his completed work could at least reach a general audience.[27] The book grew naturally out of her own work in the Education Department of the AMNH and kept in line with her husband's commitment to his curatorial duties at the museum: to maintain the collection, conduct original research, and ensure that the results of that research were communicated to the general public through the museum's exhibitions, radio shows, and public lectures.[28]

Noble cannot easily be categorized as either a naturalist or a laboratory-centered biologist; he represents what we might call an early organismal biologist.[29] He used experiment and observation in nature and in his laboratory to understand both the physiological and evolutionary causes of reproductive behavior, and he hoped to determine the evolutionary pattern of behavioral development in vertebrates.

Natural Reproductive Behavior in Animals

The study of natural reproductive behavior formed a crucial component of Noble's research agenda and was key to the position of his Laboratory of Experimental Biology in a museum of natural history. For Noble, his laboratories provided space in which he could investigate the normal process of evolution. Not only did he use wild-caught experimental subjects, but he also created spaces for his experimental subjects that mimicked their wild environments. He ensured that readers of his papers saw his laboratories as extensions of nature.[30] Studying wild organisms in a naturalistic, yet controlled, environment allowed Noble to unite different ways of knowing biological phenomena into a single vision of the organism in question. He considered this important because each species had a unique repertoire of sexual and social behavior—just because he could demonstrate female choice or sexual selection in one species did not mean that it was an important evolutionary factor in all species. He then used these data to track the evolutionary pattern of reproductive behavior.

For Noble, a natural research environment and wild organisms were important for several reasons. Until he could breed organisms in the laboratory, the success of his experiments depended critically on expeditions to the field to gather his animal subjects, a very common practice at the time.[31] Noble devoted a certain portion of his budget each year to collecting fishes, frogs, turtles, and birds in the natural areas of Long Island and New Jersey.[32] He also worried that the rats and flies typically used in laboratory experiments did not reflect the diversity of behavioral mechanisms that existed in nature. After countless generations in a laboratory environment, the animals probably adapted to these artificial conditions, Noble reasoned. As a result, he never tried to standardize his laboratory animals for greater experimental control. Further, the environment of the laboratory itself could induce artificial behavior in animals. Behaviors never observed in nature might be produced in a laboratory environment by a variety of unnatural factors, including a higher than usual density of organisms in a cage or artificial lighting. In his greenhouse, Noble constructed experimental environments that emulated the natural habitats of his research organisms.[33]

Two of Noble's investigations offer a flavor of his work to naturalize his laboratory spaces: experiments (posthumously published) on the mating behavior of the American "chameleon" (*Anolis carolinensis Voight*) and research (unpublished) on the mating behavior of box turtles (*Terrapene carolina*) at Jagger's Swamp.[34] In the first set of experiments, conducted with Bernard Greenberg, on the role of the male anole's dewlap (a brightly colored flap of skin hanging beneath its chin) in eliciting mating behavior in females, we can see the importance Noble ascribed to making his laboratory spaces appear natural to the organisms. At the beginning of the paper, Noble and Greenberg noted that they frequently observed homosexual interactions in crowded laboratory cages, but never in the field. They attributed this to the marked influence of space restrictions on social situations, and they redesigned the greenhouse to give the anoles a larger area in which to wander. "This procedure," they argued, "offered the advantages of controlled experimental tests while avoiding artificiality as much as possible."[35] Noble and Greenberg established two territories of similar quality in the experimental cages, for two males.[36] They placed one male in each territory and released a female into a box in the center of the cage. The female could then "choose" one of the males by walking along a branch to his territory. In these experiments, females responded to both the physical vigor with which the male displayed and the red color of his dewlap. To distinguish the effects of coloration on female choice, the researchers would cover the dewlap of one of the two males with green paint, or glue the dewlap to the bottom of his chin with colloidon, thereby rendering it invisible. They repeated this sequence a number of times, with inconsistent results—sometimes a male would not display properly or would hide in the greenery when they released the female. Much of the evidence they presented in the paper came in the form of descriptive accounts of individual encounters. For example, they took the time to describe one atypical mating, when "the female ran against the front of the cage and was trapped there in an attempt to penetrate the glass."[37] Despite their difficulties with the experimental design (as when one or both of the males disappeared behind foliage), they never removed the branches from the cage. Noble and Greenberg's use of greenery in the experimental cages and the greenhouse to create a natural environment for the lizards illustrates the importance they accorded to a natural background in which experimental behaviors should be measured.

Noble's framing of his research problems in terms of understanding the context of natural behavior is also illustrated in his research on the reproduction of spotted turtles. Field research began in 1936 at Jagger's Swamp on Long Island. He had already conducted research on box turtles in his laboratory, but wanted to ensure that the behaviors he had observed in the laboratory could also be observed in a natural setting. When the behaviors corresponded, Noble concluded that these were behaviors naturally exhibited by the turtles in both spaces. Problems arose when the data from laboratory and field investigations conflicted. For example, in his laboratory, Noble recorded multiple observations of homosexual behavior (he used this term to describe the act of males "allowing" other males to mount and attempt to copulate with them). Yet he did not observe any homosexual behavior at Jagger's Swamp: "our field work, however, rendered valueless these entire laboratory findings on the behavior of the ridden male, because there is no behavior of this kind among spotted turtles in their natural habitat."[38] When his data gathered in nature differed from his laboratory results, Noble made the field observations trump. Only after data from the field were gathered did Noble feel "thoroughly justified" in making conclusions about the normal breeding habits of the painted box turtles. In subsequent laboratory experiments, he further naturalized the surroundings by adding more leaves to act as ground cover, so that the females could bury themselves in the cover and the males would have to find them in order to mate.

Throughout his manuscript on turtles, Noble made reference to the congruence or difference in observed breeding patterns in the field and in the laboratory. In particular, he recorded behaviors that he deemed important in inducing the turtles to breed successfully in the laboratory, such as diurnal patterns of copulation and feeding habits. His notes also indicate that the turtles emerged from the water and sunned themselves at similar times of day in the field and in the laboratory. Noble had not included the time of sunning as an experimental factor, so it seems that he used this metric as an independent confirmation of the naturalness of the turtle's behavior patterns.[39]

While Noble clearly designed his laboratory spaces to mimic the natural surrounding of his animal subjects, he simultaneously had no qualms about performing experiments in natural habitats. In the case of the flicker, a beautiful woodpecker with a splash of red on the back of its head and brilliant yellow plumage under the wings and tail, Noble set out to understand the possible role of this coloration in its sexual and social behavior. In *Descent of Man and Selection in Relation to Sex*, Darwin had identified the flicker's mating behavior as a perfect example of female choice—several "gay suitors" courted a single female until she demonstrated a "marked preference" for one of the males. Julian Huxley suggested, however, that the flicker's distinctive coloration indicated mutual selection, as both the male and female possess the beautiful yellow feathers characteristic of the species. To decide the issue, Noble designed a series of clever experiments culminating in disguising a female flicker as a male by the addition of a dark "moustache," constructed by gluing a few neck feathers from a pileated woodpecker to her lower cheeks. The effect is easier to visualize if one thinks of muttonchops rather than a moustache, as the male flicker lacks a dark band over its beak connecting the two patches of dark feathers that comprise the "moustache."[40]

During the spring of 1935, in a willow tree in Noble's backyard in Englewood, New Jersey, a male and female flicker established a territory. Using mounted dead flickers to test the pair's reactions to intruders, Noble discovered that the female flicker would consistently attack other females, whereas the male flicker preferentially attacked other males. Although it appeared that the two flickers consistently identified the sex of the intruder, Noble suspected their reactions were artificially constrained, given that the dead birds produced no response to the attacks. He hit upon the strategy of removing the nesting female, making her look like a male, reintroducing her to the willow, and observing the subsequent interactions. Would the male be able to recognize his partner? The answer was no. When the newly disguised female alighted on the willow near their hole, it so happened that her head was turned away from the male. He approached her from behind and began courting, but when she turned around his disposition changed dramatically. "With a start he drew back and recovering his balance, lunged viciously forward. Then began a long drive which lasted nearly two hours and a half. Along the branches, back up to the trunk, up to the nest hole and again out on the branches the male mercilessly pursued the moustached female."[41] Unsure of whether the male was confusing the disguised female for a male or was simply chasing away all "foreign" birds, Noble next conducted twenty-eight trials in which the moustached female was tethered to the willow on the far side of a stuffed, mounted female flicker. The male consistently walked around the mounted female to display in front of and then attack his former partner, at which point Noble would frighten him away before the attacks caused any lasting harm.

The lesson for Noble was that through careful manipulation, he could differentiate between the effects of social behavior (natural selection) and sexual behavior (sexual selection) in determining the interactions of his animal subjects. He concluded that the "dance" of the flicker did not stimulate mate choice, but instead was involved in territory demarcation and defense. Similarly, neither female choice nor mutual selection played a role in evolution of the bright yellow coloration. Instead, Noble identified the dark moustache of the flicker as the characteristic "badge of maleness." Yet even this dimorphism, he argued, acted more as a stimulus for territorial disputes than as sexual display.

Noble carefully recounted each step of his experiments with the flickers. By adopting this narrative style, he emphasized the unique behaviors of each species he investigated, as well as individual variations within a species. This allowed him to compare social and sexual behavior across a wide variety of animals. Noble became convinced that birds, for example, exhibited a wider array of courtship behavior than did fishes. In some bird species, including the flicker, the female invited the male to mate by using a distinctive cry; without such a cry, the males did not attempt to copulate with the females.[42] Noble never observed any such invitation behaviors in fishes. As a result of his observations on birds, he suggested that bird courtship was more "stimulative" in nature, while mammalian courtship was more "emotional."[43] Despite such variation in sexual behaviors, however, Noble insisted that all vertebrate species exhibited some form of social dominance behavior. By describing the behaviors he had observed in each of these large taxonomic groups, Noble could discuss the evolution of social and sexual behavior across wide swaths of the animal kingdom without extrapolating his observations from one level of biological organization to another.[44] Although he suggested that whenever multiple members of the opposite sex were present, some form of mate "choice" must be occurring, species differed in the specific manifestations of courtship behavior.[45]

When Noble returned to his laboratory work, he continued to consistently obtain experimental subjects from his field sites. In his view, these organisms, gathered "in the wild," stayed wild in his laboratory because the conditions of his experimental cages re-created their natural environment. For example, in describing his laboratory in a funding letter to the NRC-CRPS in 1935, Noble described it as "unique in having been built for the purposes of studying the behavior of a variety of wild forms under ideally controlled conditions."[46] He kept his experimental aquaria and cages sufficiently natural to ensure that the behaviors of his wild-caught subjects were not influenced by their new, artificial environments.[47] Even so, the next year, in a similar letter requesting a continuation of funding from the NRC-CRPS, he noted that "considerable time was spent in the field this spring in order to prove that the social systems worked out with the caged animals actually exist in nature."[48]

Thus, Noble worked to make his laboratory space at the AMNH an extension of his work in fields and bogs surrounding New York. The laboratory facilities gave him greater ability to identify and understand the reproductive behavior characteristic of many animal species. He then used this understanding to begin reconstructing the evolution of social behavior in vertebrates.[49] Noble carefully regulated the naturalness of the experimental spaces in his Laboratory of Experimental Biology as a method of gaining access to the underlying mechanisms of the hormonal control of normal animal behavior. Through his experimental manipulations in both the field and the laboratory, he painted a picture of mating behavior in vertebrates that included an important space for female choice in some, but not all, of the species he investigated. Noble's findings described a fairly straightforward hierarchy of true "choice" behaviors as characteristic of species with more "complex" neural functions and more intricate courtship behaviors.

Laboratory Transitions:
From Experimental Biology to Animal Behavior

Following Noble's death in December 1940, the future of the Laboratory of Experimental Biology was uncertain without the force of his personality and focus of vision. The AMNH suffered financially during the Great Depression, and at least one trustee urged the museum to abandon the LEB and donate the facilities to a nearby university. In 1941 alone, the trustee argued, this would represent a $10,000 saving in operating costs.[50] Yet the laboratory survived. One of the lasting effects of the ambiguity with which the museum regarded the laboratory was the need for its curators to justify their research in terms of the museum's goals.

When Noble died, Roy Chapman Andrews, director of the AMNH, threatened to close the laboratory. Young psychobiologist Frank Ambrose Beach was placed temporarily in charge of the department.[51] The future of the laboratory was further complicated by other circumstances; in the summer after Beach assumed his position, and while he was doing field research in the Rockies, all of the WPA workers at the museum were dismissed. This drastically reduced the available workforce and resulted in the death of many of the organisms kept in the laboratory, as no staff remained to care for them.[52] When Andrews asked Beach how long it would take to finish the ongoing research, Beach interpreted the question as a threat to terminate the research program, but he refused to "roll over and play dead." Calling on the biggest guns he could muster, Beach wrote to Karl Lashley, a long-standing member of the NRC-CRPS and one of Beach's dearest mentors, and Robert M. Yerkes, director of the NRC-CRPS, psychology professor at Yale, and director of the Orange Park Primate Laboratory in Florida, entreating both men to talk to Andrews.[53] W. Douglas Burden, a major contributor to the museum and an old friend of Noble, also wrote convincingly to the trustees.[54] Through their efforts, the department survived and Beach was promoted to curator.

Beach followed in Noble's footsteps by focusing the laboratory's resources on the experimental investigation of comparative animal behavior. Like Noble, Beach argued for including a variety of experimental organisms in animal behavior studies, though he did so for different reasons. Trained in experimental psychology

"Are we building a general science of behavior or merely a science of rat learning?" Frank Beach worried about the siren call of the Norway rat and its effect on American experimental psychologists. He advocated using a variety of animals to model different aspects of human behavior. From Frank A. Beach, "The Snark Was a Boojum," *American Psychologist* 5 (1950): 115–24, fig. 3, published by the APA; used with permission.

by Lashley (and others), Beach wished to infuse the community of experimental psychologists with the comparative approach he had acquired at the AMNH. He criticized American psychologists for their reliance on the rat as their sole model organism. In his 1949 address to the Division of Experimental Psychology of the American Psychological Association, Beach asked: "Are we building a general science of behavior or merely a science of rat learning?"[55] Although he agreed that rat mazes were useful for understanding trial-and-error learning in humans, they were useless, he argued, for understanding the role of reasoning or insight in learning— that is, the "upper limits of intelligent behavior in mammals or other highly developed animals."[56] Beach argued that different animals were suited to different kinds of research questions—for example, he came to believe that dogs were the best organisms with which to model the reproductive behavior of humans.[57] Whereas

Noble included multiple species in his research on the evolutionary development of social behavior, Beach included multiple species to model and understand different aspects of human behavior.

Among his colleagues, Beach developed a reputation of being extremely open-minded about scientific theories, and he was one of the key figures in importing ethological ideas into the United States after the Second World War. Taking a cue from ethologist Konrad Lorenz and from his former advisor, psychologist Karl Lashley, Beach dismissed as anthropomorphic the study of individual motivation as the basis of behavior, along with feelings and emotions. Instead he argued that in higher mammals, such as humans and other primates, control of sexual behavior stemmed from the cortex of the brain, leading to a greater reliance on social learning in the development of sexuality than in the lower mammals, which therefore exhibited a smaller repertoire of sexual behaviors.[58]

Although Beach's position at the AMNH allowed him plenty of time for research, he chafed at his administrative duties. In 1946, he jumped at the chance of a faculty position in psychology at Yale University's School of Medicine, where he would teach only upper-level seminars. Lester Aronson took Beach's place as curator, albeit as head of the Department of Animal Behavior, the new name Beach had bestowed several years earlier.[59]

The question of the laboratory's fate again arose, and the burden of proving that the experimental study of animal behavior fit within the larger goals of the AMNH now fell on Aronson's shoulders. On the one hand, Aronson's research agenda directly echoed Noble's (rather than Beach's) commitment—to comparative study: "Our main goal is not to apply the answers to human behavior directly, but rather to improve our understanding of the evolution of behavior."[60] On the other hand, Aronson's experiments differed significantly from Noble's emphasis on the pattern of behavioral development; Aronson sought to demonstrate how behavior could affect the process of evolution. The disparity between Noble's and Aronson's attempts to unite behavior and evolution into a single explanatory framework reflected the rapidly changing emphasis of contemporary evolutionary theory.

Aronson's experiments on the role of behavior in determining the reproductive isolation of natural species fit neatly with the research agendas of curators from other departments at the AMNH. For example, E. Thomas Gilliard, curator of the Department of Ornithology, investigated the possible role of display and song in maintaining species isolation among the birds of paradise and bower birds of New Guinea.[61] Charles Bogert, Noble's successor as curator of herpetology, published on the role of mating calls as a reproductive isolating mechanism in toads.[62] In fact, research on reproductive choice as an isolating mechanism formed one of the central questions in the study of animal behavior in the United States, starting in the mid twentieth century.[63]

During the 1940s, the study of animal behavior in an evolutionary context at the AMNH was substantially transformed, in terms of both personnel and methodology. It is to Aronson's efforts at fitting the laboratory's research into the agenda of the museum that we now turn.

Reproductive Behavior and Isolating Mechanisms in Guppies

Lester Aronson's interest in reproductive behavior as a mechanism for speciation was piqued through a unique confluence of biologists working at and associated with the AMNH. From 1942 to 1968, the experimental tanks of the New York Aquarium were housed at the AMNH—first in the Laboratory of Animal Behavior greenhouse (in space vacated when most of the animals died after departure of the WPA workers) and then on the sixth floor of the Whitney Memorial Wing (the ornithological section) of the museum.[64] The curator of the New York Aquarium, Myron Gordon, brought with him a research focus on the genetics of cancer and coloration in fishes.[65] Charles Breder became curator of ichthyology in 1944 and maintained a research interest in the evolution of reproductive modes in fishes throughout his career.[66] When Eugenie Clark, a graduate student at New York University, came to the AMNH in 1948 to conduct her doctoral dissertation research, her interests focused the attention of these men on the question of how reproductive behavior could affect the genetics of speciation in fishes, especially guppies.[67]

Guppies exhibit a unique set of reproductive behaviors that make them easy and interesting to observe. They are internally fertilized. The male possesses a modified anal fin, called a gonopodium, that he uses to stuff his sperm into eligible females. The eggs then develop inside the female, although they do not receive any nutrition from the mother, and eventually she "gives birth" to the young after they have hatched internally. Guppies do not generally hybridize in nature, but this reproductive isolation between species often breaks down once they are placed in aquaria. As a result, guppies had been popular subjects for both scientific researchers and home aquarists for almost a century.[68]

Caryl Haskins and Edna Haskins, at New York's Union College, provided an additional spark to Aronson's curiosity about the role of female preference in the reproductive isolation of guppies.[69] Because male guppies of different species vary strikingly in shape and coloration (and differ from females in both appearance and behavior), Haskins and Haskins hoped to demonstrate that female mate choice played an important role in the maintenance of reproductive isolation between species. In their initial experiments, they were delighted to find that sexual selection did seem to be a factor—albeit in a modified form. In two different experimental setups, they presented a group of three females (one from each species) with numerous males of other species, and presented a single male with two females of different species. Haskins and Haskins observed the total number of "gonopodial contacts" between males and females (attempted copulations on the part of the male) and noted the species identity or identities of the pair. If gonopodial contacts occurred more frequently between males and females of the same species than between different species, then sexual selection would be maintaining reproductive isolation between the species. Although the researchers noted that the males seemed to mate randomly with the females at first, after a stabilization period of at least a week, the males' powers of discrimination improved remarkably. In their conclusion, the Haskins remarked that "discrimination is made on the part of the male alone, the female being an essentially passive agent in the process."[70] If sexual selection were occurring, it appeared to be through male choice rather than female choice—an observation, they argued, that did not sit well with the fact that the males were larger and brighter than the females.

These experiments by Haskins and Haskins also illustrated the continuing effort biologists expended to modify artificial laboratory environments to mimic nature. They arranged a fifteen-gallon aquarium "to simulate as closely as possible a section of the lagoon environment. This arrangement was made with considerable care since it was deemed important to maintain the ecological situation as nearly intact as possible."[71] Wherever possible, they also used wild-caught guppies. As such, they tried to re-create a natural environment in which to observe the normal behavior of their subjects.

Following on Haskins and Haskins's research, when Aronson began collaborating with Gordon and Clark, the three colleagues chose to investigate the importance of psychological isolating mechanisms in two different species of Mexican guppies: the platyfish (*Platypoecilus maculatus*) and the swordtail (*Xiphophorus helleri*). Clark, Aronson, and Gordon intended their research on guppies to throw light on species-isolating mechanisms generally. Their early investigations revealed that although males frequently initiated copulations (gonopodial "jabs" at female genitalia lasting a fraction of a second), only a small percentage maintained genital contact for as long as a couple of seconds, the time required to successfully transmit sperm.[72] So, although it initially seemed that mating took place almost at random in the tank, actual copulations were much more selective. Clark's experiments with artificial insemination in these species also indicated that even if behavioral isolating factors broke down, species isolation was not necessarily lost. She suggested that if a male were to stuff his sperm inside a female of a different species, as long as the female's genital duct also contained sperm from a male of her own species, competition between the sperm of the two males would lead to the species-appropriate sperm fertilizing her eggs. Thus, the complete reproductive isolation observed in nature, they concluded, was due to the cumulative effect of several partial isolating factors—psychological (behavioral incompatibility), ecological (habitat separation), morphological (making genital-genital contact between different species difficult or impossible), and physiological (sperm competition).[73]

To determine the relative importance of mating behavior to other isolating factors in preventing hybridization between these species of guppies, Clark, Aronson, and Gordon conducted a series of "male-choice" tests. In a single aquarium, they placed one male and two females (one of the same species as the male, one of a different species). They then recorded which female the male mated with. It is possible they chose to use "male-choice" tests rather than "female-choice" tests because the latter experiments in guppies seemed technically infeasible. Males interrupt all rivals' attempts at courting available females, with the result that it can be impossible to document a successful courtship attempt when more than one male is present in the tank.[74]

In spirit, Clark, Aronson, and Gordon were continuing Noble's investigations of sexual selection in fishes, while in practice their research differed considerably. In their joint paper, they included forty-eight tables of data, summarizing more than seventeen hundred ten-minute observations of mate choice—a far cry from Noble's narrative descriptions of a few experimental trials. Clark had started recording all their observations by hand, using a separate mark for each kind of behavior, but she found it necessary to take her eyes off the fish in order to jot down comments or to look at the clock. To solve this problem, they rigged a typewriter with a polygraph to make a machine that allowed them to hit particular keys for certain behaviors, markedly increasing the efficiency of the observer. Further, Clark, Aronson, and Gordon did not self-consciously attempt to emulate the natural environment in their experimental spaces in the same way as had Noble. They removed all plants from the observation tanks (although not from the tanks where the fishes typically lived), because the plants sometimes obscured the courting fish from the view of the other fish, thereby delaying courtship activities, or prevented the researchers' ability to observe what was happening in the tank.

In the end, the three researchers found no evidence of mate choice. The male guppies would often attempt to copulate with any female with which they were presented, regardless of species. Clark, Aronson, and Gordon concluded that mate choice could not be an important isolating mechanism in guppies and therefore that behavior was not an important factor in the evolutionary process of speciation in guppies.

In an early grant application to the NRC-CRPS (the same institution that aided Noble's research during the Depression), Aronson couched their experiments on behavior as a mechanism for reproductive isolation in terms of recent research with experimental population genetics of fruit flies. He cited the research of geneticists Theodosius Dobzhansky and Herman Spieth, as well as ornithologist Ernst Mayr's work on the importance of psychological isolating mechanisms in the speciation of fly populations that were, in theory, capable of interbreeding (they did so in a laboratory environment) but did not interbreed in nature. All of these men lived and worked in the New York area; Mayr worked down the hall at the AMNH and frequently collaborated with Dobzhansky, whose research laboratory was based at Columbia University; Spieth became friends with both Mayr and Dobzhansky and, thanks to their encouragement, began to explore the mating behavior of fruit flies at the College of the City of New York, where he taught. Building on the interests of this unique New York community, Aronson framed the proposed work on fish breeding as a test of "this fundamental problem of evolution" in Mexican guppies— namely, that guppies would interbreed in domestic aquaria even though they "never hybridize under natural conditions."[75] In the published discussion of their results, Clark, Aronson, and Gordon comprehensively reviewed the literature on the role of the male and female in sexual isolation in both fishes and *Drosophila*, grounding their study of sexual selection in recent experimental population genetics.[76]

Rather than thinking about how evolution shaped the behavior of individuals, Clark, Aronson, and Gordon had tried to determine whether the behavior of individuals could change the evolutionary future of a species. This represented a distinctive shift in the relationship between behavior and evolution at the AMNH. To address why this research focus changed, we must turn our attention to the contemporary research of experimental population geneticists working with species of the fruit fly, *Drosophila*.

Fruit Flies, Reproductive Isolation, and Female Choice

Among historians of science, the greatest credit for changing the study of the evolution of natural populations of animals in the United States during the 1940s is generally accorded to two men: Theodosius Dobzhansky and Ernst Mayr.[77] In 1927, Dobzhansky left his native Soviet Union to work in Thomas Hunt Morgan's *Drosophila* laboratory. Exactly one decade later, Dobzhansky published *Genetics and the Origin of Species*.[78] In his book, Dobzhansky made available in readily accessible

Male *Drosophila melanogaster*. This species of *Drosophila* was named for the "dark belly" of the male. Similar sexually dimorphic pigmentation patterns are common in this genus. Courtesy of Benjamin Prud'homme and Nicolas Gompel.

prose the conclusions of the mathematical population geneticists of the 1920s and early 1930s: small genetic changes in a population could, over time, account for large physical differences between populations. In 1940, Dobzhansky returned to Columbia University, less than three miles from the AMNH in New York City. Mayr, a Bavarian by birth, emigrated to the United States in 1931, when he accepted a post as curator of the Rothschild ornithological collection at the AMNH. Mayr subsequently published *Systematics and the Origin of Species* (1942), in which he suggested that such processes operating in a single species (microevolution) were functionally equivalent to the processes governing speciation, when one population split into two populations (macroevolution).[79] Together, these two books provided Americans with a new research focus for understanding the processes regulating evolution in natural populations, and new ways of defining animal species.[80]

Dobzhansky and Mayr thought of species in very similar ways, as self-defining groups in which members of the population mated only with other members of the same population. In *Genetics and the Origin of Species*, Dobzhansky defined "species as that stage of evolutionary process, 'at which the once actually or potentially interbreeding array of forms becomes segregated in two or more separate arrays which are physiologically incapable of interbreeding.'"[81] Mayr modified this definition only slightly: "Species are groups of actually or potentially interbreeding natural populations, which are reproductively isolated from other such groups."[82] Their definitions of species as interbreeding populations generated many new questions, of which I will highlight two. First, how did species become reproductively isolated?

Second, was it possible to use the degree of sexual isolation between populations as a way of diagnosing species? Searching for answers to these questions formed important components of the research program of experimental population genetics in the 1950s.

Dobzhansky identified two broad categories of mechanisms that could potentially act to isolate populations genetically—geographic and physiological.[83] He defined physiological isolating mechanisms broadly and included everything from ecological specializations to morphological and behavioral differences between species and physical incompatibility of gametes. The relative importance of these physiological isolating mechanisms was a subject of debate among biologists in subsequent decades and was a question Clark, Aronson, and Gordon sought to answer with their experiments in the Laboratory of Animal Behavior at the AMNH.[84]

In 1944, Dobzhansky and Mayr proposed an experimental method for quantifying the degree of reproductive isolation between two species of *Drosophila*.[85] In a laboratory, they presented males of one species with females of the same species and females of different species. They defined the degree of reproductive isolation as the proportion of males that chose mates of the correct species—the fewer "mistakes" made by the male, the greater the proportion of intraspecific matings and the more isolated the two populations. This male-choice model became the first standard methodology for determining whether two populations were sexually isolated.[86] It was the same male-choice experimental design cited and used by Clark, Aronson, and Gordon in their experiments.

Dobzhansky and Mayr's methodology soon came under fire, however, as a few biologists investigating the process of speciation began to claim that female preference was more important than male preference in maintaining reproductive isolation between closely related species. The spark for this reaction against the Dobzhansky-Mayr male-choice experimental model seems to have been a paper by J. M. Rendel published in 1945.[87] In several breeding experiments, Rendel observed that males showed no obvious preference for either kind of female (that is, of the same or different species), while females seemed to exhibit a preference for conspecific males. Rendel believed that the apparently greater ability of females to differentiate between males was actually the result of species-specific stimulatory behavior. Males of one species were not capable of stimulating a female of a different species sufficiently to induce a mating response. Therefore, Rendel postulated, female resistance to males of inappropriate species might act as a mechanism for speciation.[88]

A second paper also helped establish that females, rather than males, were the primary determinants of reproductive isolation. In 1948, George Streisinger, a geneticist who started his research while spending the summer at Cold Spring Harbor, presented male fruit flies with their choice of etherized (unconscious) or non-etherized (awake) females of several species.[89] Streisinger noticed that when the females were awake, all copulations involved males and females of the same species. When he etherized the females, effectively knocking them unconscious, the males mated with all females indiscriminately. "No choice is discernible when etherized females are used," Streisinger remarked dryly. He also noted that with etherized females, males of one species in particular, *Drosophila pseudoobscura*, were far more likely to court each other than to pay attention to the motionless females. Based on this evidence, Streisinger suggested two possible explanations. Perhaps males were exhibiting species-specific behavioral signals to which females needed to respond for "correct" pairing to occur. Or, females might provide species-specific stimuli, whereupon males responded either by completing the copulation or by dismounting. In either case, Streisinger maintained, the role of females was crucial to maintaining reproductive isolation between these species of fruit fly.

Rendel's and Streisinger's experiments convinced English geneticist Angus John Bateman that the topic merited further investigation.[90] At the John Innes Horticultural Institute, Bateman worked principally on population genetics and mutagenesis in plants, foraying occasionally into animal genetics.[91] In 1948 and 1949, he published two papers on mate choice in *Drosophila*, papers that inspired Ronald Fisher's scorn. In addition to giving a male multiple active females among which to choose, Bateman also gave one female several kinds of males. He used fruit flies that exhibited particular traits that would always be expressed in their offspring, or "marker traits." He tracked which copulations produced offspring. Although it is generally easy to identify the mother of young flies (she's the one that deposits the eggs), it can be impossible to identify the father; thus, biologists often used mating success as a proxy for reproductive fitness. For example, Rendel and Streisinger both relied on the presence of sperm in the female as a sign that the male had succeeded. Bateman, however, actually measured which males produced the highest number of total offspring in each of his experimental treatments. This was no mean feat; it was possible only because sufficient marker traits had been identified in *Drosophila*. From his experiments, Bateman concluded that Rendel's and Streisinger's suggestions were correct: females were more likely to exhibit species-specific mating preferences than males. He then enjoined other drosophilists to use female-choice experimental tests to determine the degree of reproductive isolation between two populations, rather than using Dobzhansky and Mayr's male-choice tests.[92]

Bateman further argued that sex differences in flies' ability to differentiate between partners of the same species and those of different species occurred because females were evolutionarily selected to be more "choosy." He noted, on the one hand, that the number of offspring sired by males increased linearly with the number of females with which they successfully mated. On the other hand, females produced the same number of offspring regardless of how many times they mated.[93] As a population geneticist, Bateman assumed this meant that females only mated once in a breeding season (in fruit flies, once in their short lives)—he saw no evolutionary advantage if females mated multiple times, because their total reproductive contribution to the next generation never varied. Because of his assumption, Bateman argued that if a male made a single mistake, it did not significantly diminish his reproductive success, but a similar mistake on the part of a female might cost her all of her offspring.[94] Females, Bateman concluded, were far more likely to be choosy than males in mate selection, as Darwin and Fisher had previously noted.[95]

Slowly, other drosophilists also began to accept the idea that females, not males, exerted primary control over the degree of sexual isolation between two populations.[96] In 1950, for example, David Merrell, an evolutionary biologist at the University of Minnesota, began a decade-long campaign advocating the use of female-choice or female-multiple-choice experiments over male-choice experiments.[97] Like Bateman, Merrell argued that sexual isolation was less likely to be demonstrated if scientists continued to use only male-choice experimental models. Through the efforts of Rendel, Streisinger, Bateman, Merrell, and others, population geneticists began to design experiments mindful that females were more likely to exhibit reproductively isolating mating behaviors, and the Dobzhansky-Mayr male-choice model began to disappear in favor of either female-choice or female-multiple-choice experimental setups.

In one sense, these experiments approached the question of behavior and evolution from a radically different perspective than had Noble's experiments on behavior in turtles, lizards, and fishes. Rather than watching individual courtship, geneticists were interested in the statistical outcome of many attempted copulations. In another sense, the experiments echoed the same concerns that Noble expressed in his double-edged attitude toward laboratory spaces. Some population geneticists of the same era performed experiments on wild-caught individuals of *Drosophila* as a way of investigating the genetics of natural populations. Dobzhansky, for example, remarked that "finding mutations *in nature* was something quite different from finding mutations produced by x-rays or something else. It was finding the existence of mutants in natural populations that was regarded as a test of their theory. Ten times as many mutations induced by x-rays would not do."[98] Such a notion of "natural" differed strongly from Noble's desire to maintain a natural laboratory environment by adding branches and leaves to his aquaria as a mechanism of ensuring normal sexual behavior of his observational subjects. In the 1930s and 1940s, biologists used their studies of reproductive behavior in both laboratory and field environments to address the goal of understanding the natural behavior of wild organisms.[99]

In the decade that followed, this new approach to behavior and evolution took a more prominent role in the research on behavior in an evolutionary context at the American Museum of Natural History, although much of this research was conducted outside the Department of Animal Behavior.[100] Eugenie Clark moved to Florida in 1955 to become director of the Cape Haze Marine Laboratory, Myron Gordon died in 1959; and Lester Aronson's attention turned to the neural regulation of sexual behavior in animals.[101] However temporarily, the issue of reproductive isolating mechanisms had provided an opportunity for curators of the Department of Animal Behavior to contribute to the museum's new research focus on the process of evolution.

In the United States from the 1930s through the 1950s, the study of mating behavior within an evolutionary context underwent drastic methodological and theoretical changes. In the 1930s, Gladwyn Kingsley Noble designed experiments to help reconstruct an evolutionary history of behavior, mapping the increasing complexity of social and sexual behavior in animals. He hoped his descriptive analyses of individual courtship would uncover the zoological antecedents of human behavior. By the 1950s, Lester Aronson designed experiments that emphasized massive quantities of experimental data and examined the outcome of hundreds of mating pairs of fishes. His collaborators hoped to prove that males of one species could differentiate between females of their own and other species.

These changes can best be described as simultaneous shifts in experimental design, theoretical framework, and underlying assumptions about the relationship of animal behavior and evolution. In terms of experimental design, Noble employed narrative explanations of mating behavior in a few individuals, whereas Aronson, Eugenie Clark, and Myron Gordon collected quantifiable statistical data on the outcomes of matings. In terms of theoretical framework, the contrast between Noble's and Aronson's research reveals a trend away from uncovering the evolution of behavior and toward understanding the process of evolutionary change. To Noble in the 1930s, behavior was a biological trait shaped by evolution; Aronson's research group investigated behavior as a mechanism of evolution. Underlying these two research programs were different theoretical approaches to synthesizing research on animal behavior and evolutionary theory. Whereas Noble incorporated evolution into an animal behavior framework, Aronson incorporated animal behavior into an evolutionary framework.

During this same period, "the experimental organism from the naturalist's point of view" also changed. Noble and Aronson fully recognized the potential influence of a laboratory environment on the expression of normal reproductive behavior in their experimental subjects. Noble sought to minimize these effects by creating laboratory conditions that reproduced the most critical natural conditions of his experimental organisms. Aronson and his collaborators took advantage of the discrepancy between natural and laboratory reproductive behaviors of guppies to investigate the relative importance of barriers to hybridization in maintaining reproductive isolation. Aronson's work of the 1950s should be seen as a combination of an interest in laboratory investigations of social behavior and a desire to study the behavior of natural organisms in an evolutionary context.[102] From the 1930s through the 1950s, zoologists at the AMNH with an interest in animal behavior sought to understand and control for the ways in which the laboratory might induce artificial behaviors.

The history of the Laboratory of Animal Behavior at the AMNH also provides us with a microcosm in which to understand the history of research on behavior and evolution in the United States more generally. Even after the mathematization of evolutionary theory in the 1920s and early 1930s, biologists interested in the evolution of behavior continued to conceive of evolution as a primarily linear process by which they could map the increasing complexity of behavior in animals. Noble certainly envisioned evolution as the history of what had happened to behavior over millions of years, and he used extant species as placeholders for the behavior of animal antecedents of human behavior. Yet, as the study of evolution in the United States shifted from mapping the evolution of particular traits or behaviors to studying mechanisms of the evolutionary process, it became increasingly unacceptable for experimentalists to study the progressive evolution of behavior, as Noble had envisioned. By the late 1940s, Clark, Aronson, and Gordon's investigations of the role of behavior as an isolating mechanism in fishes were modeled explicitly on similar experiments in *Drosophila* designed to elucidate mechanisms of speciation. The population approach to research on the genetics of *Drosophila* was remarkable for its pervasive tendency to explicitly establish what should qualify as an appropriate evolutionary question, even in the realm of behavior. Citing the direct influence of Dobzhansky's and Mayr's research on their experimental questions, Aronson and his collaborators instead began to investigate evolution as a process, and behavior as a possible mechanism for evolutionary change in nature.

The effect of the research program promoted by Theodosius Dobzhansky and Ernst Mayr on the study of mating behavior was profound. The research program of the evolutionary synthesis thus served to define and restrict how biologists investigated behavior as a function of evolutionary theory, even in the 1950s.

Shortly after the Second World War, a growing community of American zoologists stopped using animals as models to understand human social and sexual behavior. Using animals as models to elaborate a progressive, hierarchical evolutionary pattern of behavior ("from fish to man") simply did not fit within the synthesis program of understanding the process of speciation. Synthesis architects presumed that the process of speciation was similar across all animal taxa and that speciation in human ancestors occurred through the same general rules of evolution that affected speciation in flies. The evolutionary ladder had given way to a branching evolutionary tree, in which the tips of all branches represented equally "evolved" species. As a result, the behavior of nonhuman animals, at the tips of their own branches, could not be understood as simplified precursors of complex human social interactions, as stand-ins for common ancestors located deeper in the branching system of the tree.

Biologists at the AMNH also deemphasized rational choice-based behavior in animals as they switched from looking at female mating preferences among males of the same species to seeking evidence of females' ability to distinguish between males of their own species and males of other species. Whereas females' comparison of conspecific males implied a sense of cognition, biologists associated interspecific mate choice with instinctual behavior. In part, this partitioning stemmed from the conviction that interspecific mate choice should count as natural, not sexual, selection. These changes reflected biologists' larger concerns with how best to study evolution as a process and their desire to see evolution as a branching process in which all species—animal, human, plant—were equally "evolved," such that animal behaviors could no longer stand in as self-evident evolutionary precursors to human behavior.

The work in zoology in Europe using animals as direct models for human behavior in psychology continued apace, however. In the following decades, the study of animal behavior as an independent discipline began to flourish both in Europe and in the United States, within ethology and population genetics, respectively.

CHAPTER 4

Courtly Behavior

The Rituals of British Zoologists

To get light on the behavior of man, particularly his innate drives and conflicts, it is often helpful to study the elements of behavior in a simple animal.
—Nikolaas Tinbergen, "Curious Behavior of the Stickleback," 1952

In the United Kingdom, the community of zoologists interested in animal behavior grew quickly in the decades following the Second World War. British zoologists strove to produce a robust view of the role of behavior in the biology of organisms, combining research on systematics, ecology, evolutionary processes, genetics, and physiology. In an era of renewed hope for international scientific cooperation in the name of peace, and as intercontinental travel became cheaper and faster, biologists from both sides of the Atlantic Ocean met to discuss the behavior of animals. Although female choice and sexual selection remained at the fringes of the research program of most British biologists, this interdisciplinary community was crucial in providing biologists interested in courtship behavior with locations and reasons to gather together—from annual meetings, conferences honoring a retiring member of a discipline, and journals in which to publish research to graduate programs where students could earn a degree or simply spend a year learning new research techniques. At these venues, it seemed, two distinct groups of scientists found little common ground on how best to study the behavior of animals. Whereas the mostly American comparative psychologists emphasized using animals as research tools to address general psychological problems, the predominantly European zoologically trained ethologists prioritized the study of naturally occurring behavior patterns characteristic of particular species.[1]

British zoologists were diverse in their choice of experimental designs and the animals they studied, but united in their quest to understand the biological basis of natural behavior in animals—even if this sometimes meant raising them under artificial conditions.[2] In fact, the wide diversity of methodological approaches used in the postwar decades by scientists who thought of themselves as ethologists, and their continuing dialogue with zoologists who did not consider themselves ethologists, makes it difficult to identify a coherent group of scientists who self-identified as ethologists and were accepted as such by their peers. Perhaps what is most amazing about this period is that despite fundamental disagreements, zoologists interested in studying animal behavior maintained extended conversations about how best to study the behavior of animals and how, if at all, this research could be used to understand human behavior.[3]

Institutionally, ethology came into its own in the 1950s, after Dutch ethologist Nikolaas Tinbergen secured a post at Oxford University and Austrian Konrad Lorenz became director of the Max Planck Institute for Behavioral Physiology, moving into new facilities at Seewiesen, Austria, in 1958.[4] Tinbergen's arrival in Oxford may have marked the institutional commencement of the ethological community in England, but the English intellectual soil was already fertile with the behavioral research of Edmund Selous, Henry Eliot Howard, Frederick B. Kirkman, and Julian Huxley in earlier decades.[5] Known to his friends and colleagues as Niko, Tinbergen served as a focal point of the growing community of scientists interested in the scientific study of animal behavior. Tinbergen self-consciously played the role of community builder and teacher, and thanks to his international and interdisciplinary connections, the new center for ethology he established in Oxford was far from isolated or programmatic.[6] As a young man studying animal behavior in his native Holland, Tinbergen had been profoundly influenced by his friendship with ethologist Konrad Lorenz.[7] Tinbergen was also familiar with contemporaneous attempts in the United States to study the natural behavior of organisms—specifically, with the Laboratory of Experimental Biology (later, the Department of Animal Behavior) founded by Gladwyn Kingsley Noble at the American Museum of Natural History (AMNH) in New York City. Tinbergen's continuing contact with the AMNH after the war proved important in spreading ethological ideas across the Atlantic Ocean.

Impressed by the theoretical framework developed by Lorenz in tackling the subject of instinct, Tinbergen brought to Oxford experimental expertise and the sensitivities of a researcher devoted to field studies of animal behavior.[8] Lorenz conceived of instincts as innate behaviors, as behaviors "which animals of a given species 'have got' exactly in the same manner as they 'have got' claws or teeth of a definite morphological structure."[9] Such behaviors, he believed, were genetically determined and particulate in nature. Lorenz developed his hydro-mechanical model to explain the expression of such innate behaviors in the life of an organism. According to his

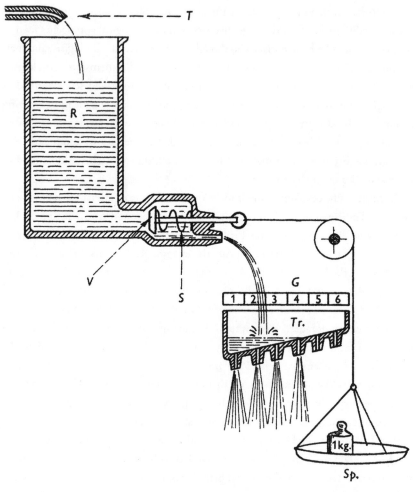

Konrad Lorenz's hydro-mechanical model of behavioral production. Only after sufficient accumulation of a behavioral stimulus (*Sp*, lower right) would the inherent drive of the animal (*R*) be released (*V*, *S*) as a behavior (the trickling water, *Tr*). Stronger stimuli and/or the length of time since the organism's last behavioral release determine the intensity of behavior produced (*G*, 1 through 6). From Konrad Lorenz, "The Comparative Method of Studying Innate Behaviour Patterns," *Symposia of the Society for Experimental Biology* 4, Physiological Mechanisms in Animal Behaviour (1950): 256, published by the Society for Experimental Biology.

model, each animal produced a certain amount of energy that caused it to want to act. The buildup of this energy, or motivation to act, would typically be released by external stimulation acting on the animal's perceptual apparatus. The behavioral act itself would consume the pent-up internal energy of the organism. If an animal failed to encounter an appropriate releasing mechanism, then the energy, motivation, or drive to act would build up to such an extent that the energy could "spark over" to a different kind of irrelevant behavior, called "displacement activities." Rather than mating, a chicken might peck at the ground; rather than fighting, a man might scratch himself behind the ear.[10] The expression of innate behavior, then, depended not only on the internal state of the animal but also on the environmental or social conditions in which the animal found itself.

By 1960, the theoretical foundations that epitomized the writings of Lorenz and Tinbergen before the Second World War were already beginning to crumble, from the distinction between "innate" and "learned" to the importance of "drive" in producing behavior.[11] Yet British zoologists continued to attempt to synthesize into a single framework their research on a rich diversity of organisms. In an article published in *Scientific American*, Tinbergen declared: "I believe that one should not confine one's work entirely to a single species. No one who does can wholly avoid thinking that his animal is The Animal, the perfect representation of the whole animal kingdom."[12] Such a statement acted both to criticize American comparative psychologists' obsession with rat behavior and to defend ethologists' interests in investigating behavior in a wide variety of species.

Certainly, bower birds and fruit flies provided qualitatively distinct examples of how courtship behavior could be produced in organisms and the function of courtship in the life of the animal. Bower birds had been emblematic of sexual selection ever since Darwin's illustrative use of their mating arenas in *Descent of Man and Selection in Relation to Sex*.[13] Biologists had never suggested, however, that bower birds' courtship behavior was typical or representative of courtship in other birds or animals; bower birds are exceptional because of their exaggerated nest-building and decorating behavior. In the 1950s, Alan John Marshall's behavioral research on bower birds centered instead on how the birds' internal environment, or physiology, interacted with external, ecological stimuli to produce behavior. Marshall cast bower birds' mating choices as the result of differential male stimulation, not female choice. *Drosophila* also constituted an important research organism for ethologists, despite our typical association of the standardized fruit fly with genetics research. Starting in the 1930s, fruit flies became the quintessential research organisms of experimental evolution—they were considered neurologically simple enough to behave naturally in a laboratory setting and to serve as a legitimate model of evolutionary processes in nature. In the 1950s and 1960s, for two of Tinbergen's graduate students at Oxford, Margaret Bastock and Aubrey Manning, fruit flies offered a convenient organism in which to study the evolution and genetic basis of innate courtship behavior.[14]

This nexus of postwar behavioral science would seem, at first glance, to be a likely place to find research on female mate choice. Controversial population geneticist J. B. S. Haldane, and his wife, the geneticist Helen Spurway, once remarked that ethologists had a narrow-minded obsession with sex. There were four behaviors necessary for the propagation of a species, "breathing, drinking, eating, and reproduction, often in that order of priority. Ethologists have studied them in the opposite order of priority."[15] Ethologists found that ritualized, innate behaviors were predictably easy to observe during courtship and much harder to observe during other kinds of organismal activity.[16] Yet female choice played only a minor role in ethological theories of behavior.

Zoologists within the ethological network generally ignored female choice as a mechanism of evolution, for a variety of reasons. The theoretical models of behavioral production, which ethologists both used and criticized, left little room for "choice" as a cause of behavior in animals. Additionally, ethologists studied the action of the evolutionary process in modifying animal behavior, rather than studying how behavior could act to alter the process of evolution (as were contemporary biologists at the AMNH). Zoologists further argued that courtship differences among species could have arisen without selection for divergence, because the major characteristics of courtship (aggressiveness, timidity, color, sexual dimorphism) are functionally important for other kinds of behaviors, too.[17] In other words, British zoologists emphasized how and why reproductive behavior varied among species, not how reproductive behavior could potentially drive the evolution of sex-based differences in appearance. Even when Tinbergen attempted to unify the increasingly disparate approaches employed by ethologists with his "four questions" for studying behavior in animals, he did not create space for behavior as a cause of evolutionary change in a species. What I want to emphasize is that British zoologists did not reject female choice per se; it just didn't fit within the explanatory frameworks they used to discuss behavior. As a result, British zoologists reconfigured the functional causes of courtship behavior without recourse to conscious choice on the part of their animal subjects. When a few ethologically trained zoologists turned their attention to humans, they applied the same kind of methods to an analysis of human courtship and once again concentrated on male-male aggression as a structuring element of social interactions, to the exclusion of female choice.[18]

Tinbergen's Four Questions

As part of his vision for the future of British ethology, Tinbergen sought to distance himself from what he considered typical behavioral research in the United States.[19] To Tinbergen, the ethological approach was superior to a "comparative" psychological rat-maze approach, because it not only took into account a larger array of animal species and behaviors, but also endeavored to understand the organism as a whole rather than concentrating on a single aspect of the organism (learning) in isolation from its relation to its environment. Tinbergen despaired that "in the U.S. potential ethologists became behaviourists and in Britain they became ecologists."[20]

Not all American psychologists were equally anathema. Beginning with Tinbergen's connection to Noble in the 1930s, the behavioral research programs at the AMNH and at Oxford were in early and constant communication. Tinbergen first visited the American museum during the summer of 1938, on a two-month trip to the United States funded by the Dutch-American Foundation.[21] During his visit, he was impressed with Noble's attempts to study animal behavior experimentally, combining both observation of behavior in the wild and experimental manipulations in laboratories at the museum.[22] Noble was similarly excited by Tinbergen's approach to the study of behavior and, in 1939, offered him a position as the resident naturalist at the Edmund Niles Huyck Preserve in Rensselaerville, New York. Tinbergen declined, and Noble reoffered the position to him the following year.[23] Tinbergen again declined, writing now that he could not leave Holland in its time of need. "There *is* war," Tinbergen wrote to Noble, "and—to put it crude—our country can be chosen to function as a battlefield every day. As long as this acute danger exists I feel obliged to stay here, as long as our government enables me to earn a living."[24]

The working group at the AMNH was key in disseminating the work of Tinbergen, Lorenz, and other European students of behavior to an American readership. In 1937, Noble set his translators the task of making available in English the full text of Lorenz's 1935 article "Der Kumpan in der Umwelt des Vogels" ("The Companion in the Bird's World"). Around the same time, a long English-language abstract of the article was published in the *Auk*,[25] but the full version was available only through Noble. To help defray costs, Noble convinced the philanthropic Macy Foundation to mimeograph and distribute several copies of the full translation,[26] and he sent at least twenty-two additional copies to interested parties.[27] During his visit to the AMNH in 1938, Tinbergen also met an American teenager conducting research on egg-incubating behavior in the laughing gull, Danny Lehrman.[28] By the time they met again, in 1954, it was under rather different circumstances.

In 1953, Lehrman published a critique of Lorenz's concept of innate behavior.[29] Five years earlier, Lehrman had entered the Ph.D. program in psychology at New York University and started his dissertation research under the supervision of T. C. Schneirla at the AMNH.[30] Schneirla pushed Lehrman to write the critique, emphasizing the importance of "experience" in the development of behavior and criticizing what Schneirla saw as Lorenz's naive acceptance of a completely hereditarian basis for behavior.[31] In his paper, Lehrman claimed that the Lorenzian concept of innate, genetically based behaviors was too simplistic; it failed to account for how behavior developed in an individual. No behavior, he insisted, could be 100 percent innate, because the expression of any behavior depended on the circumstances of the animal, just as no behavior could be 100 percent learned, as these behaviors also depended on the inherent ability of the animal to learn. Lorenz's dichotomy of innate behavior (the result of genetics) and learned behavior (the result of experience) was misleading; all behavior developed as a result of both genetic and environmental factors.[32] Further, Lehrman continued, these difficulties were merely amplified when Lorenz or Tinbergen analogized animal and human behaviors. Drawing attention to a paper in which Tinbergen suggested that searching for a nesting site in birds was akin to searching for a house in man, Lehrman pointed out that very different factors were at work in the two cases. Nominally they might appear similar—they are both *searching* behaviors—but such a label elided fundamental differences in the perceptual and cognitive faculties brought to bear in each case, even if the outcome was similar.[33] For Lehrman and Schneirla, "instinct" as a concept was barren of intellectual merit, especially when applied to humans.

Lehrman published his attack on the ethological concept of instinct at a fortuitous time. The following year, he traveled to the United Kingdom and European ethologists crossed the Atlantic in the other direction to attend the 1954 Macy Conference on Group Processes in Ithaca, New York.[34] On both occasions, zoologists spent considerable time discussing Lehrman's critique of ethological assumptions about heredity. When Lehrman arrived in Oxford, Tinbergen's "hard core" group of graduate students was initially wary, but the students received Lehrman warmly when they realized he was a fellow birdwatcher and sympathetic to the idea of studying organisms in their natural habitats. According to Aubrey Manning, one of Tinbergen's graduate students at the time, "it rapidly became clear that he was in fact deeply sympathetic to the ethological approach and also hostile to much of white-rat experimental psychology. He wanted us to be more critical in our approach to behavioral development—that was really the crux of the matter—and who can dispute that he was absolutely correct."[35] Later that year, a host of European ethologists, including Robert Hinde, Gerard Baerends, J. J. Van Iersel, and, of course, Lorenz and Tinbergen, attended the Macy Conference on Group Processes. At this week-long conference, Lorenz felt attacked by the other participants, but Tinbergen decided to take their criticisms seriously and to include them in his subsequent theorizing about the evolution of behavior.[36]

Tinbergen incorporated Lehrman's ideas into his lecture notes the following year, and later into his published work as well.[37] In his abbreviated notes for a class lecture, Tinbergen argued that the 1954 conversations between psychologists and ethologists had been constructive in clarifying what ethologists meant when they used the term "innate." He suggested that innate differences could be demonstrated only when individuals were raised under exactly the same environmental conditions, which never happened in nature. Yet, Tinbergen continued, the behaviors of animals should be considered as characters in the same way that morphological features were characters. His solution to this dilemma: students should use the phrase "non-learnt" behaviors, meaning behaviors that animals exhibited without conditioning, without imitating other animals, without practicing the behavior, and so forth.[38] Tinbergen thought that such nonlearned behaviors were ignored by most American psychologists and thus it fell to the ethologists to explore the biological basis of these behaviors more fully.[39]

When Tinbergen formalized the ethological approach in four methodological questions, he tried to incorporate Lehrman's critique in a serious way. All behaviors, he posited, have a cause, an ontogeny, a function, and an evolutionary history; to truly understand a behavior, one must address all four levels of analysis. Tinbergen first included the four questions in his lectures in the late 1950s, and in print, he most fully elaborated these questions in his 1963 paper "On Aims and Methods of Ethology."[40]

— *Cause*: What immediate physiological or psychological mechanisms contribute to the expression of behavior?
— *Ontogeny*: What is the developmental basis of behavior? In other words, how does an animal change the way it reacts to a stimulus over the course of its lifetime?
— *Function*: How does this behavior help the animal survive and reproduce?
— *Evolution*: What was the evolutionary history of the behavior in question?

Tinbergen intended the question of a behavior's ontogeny, more so than the other questions, as a solution to Lehrman and Schneirla's critique of ethological theory.

Tinbergen's methodological encapsulation of the discipline also avoided controversies over theoretical models of behavior production and encompassed many of the research programs already advanced under the name *ethology* in an increasingly fractured discipline. Different research schools within the ethological network often addressed the four questions with varying force. Tinbergen's group at Oxford was primarily interested in the functional significance of behavior, while the Cambridge school was much stronger on development, in part because these researchers were much more interested in language and communication in animals than was the ethological community at Oxford.[41]

When talking about the function of behavior, one theory in particular, ritualization, proved especially useful for the Oxford group. For Tinbergen, *ritualization* denoted the evolutionary process by which a movement or posture became specialized as a signal with a new function.[42] Julian Huxley had found the concept of ritualization particularly helpful when he discussed the evolution of courtship behavior and territoriality. Following Huxley, Tinbergen suggested that "a great majority" of behaviors characteristic of a species had undergone a process of ritualization—that is, generalized courtship behavior had been turned into species-specific displays.[43] Ritualization was also key to Lorenz's idea that in many species, intraspecific fighting had been reduced to a "tournament" in which no individuals were fatally injured and male aggression was redirected into sexual stimulation.[44] Ritualized behaviors acted as "signals" to other individuals of the same species and induced an appropriate behavioral response. The same ritualized behavior—say, a colorful display by a male bird—might cause sex-specific responses, from aggressive displays in other males to increased sexual interest in females. In territorial animals, Lorenz suggested, courtship behavior could act to spread competing males over a larger area, thereby reducing the number of fatal encounters resulting from male-male competition for access to females. In the minds of zoologists, the concept of ritualization linked strongly the ideas of courtship and aggression.

At Oxford, zoology students analyzed the mating behavior of several animal species as a way of determining the role of ritualization in courtship more generally. They argued that studying an organism in its natural habitat was essential for understanding the specific ecological conditions that made the behavioral trait adaptive for that organism.[45] Yet not everyone agreed on this point. Tinbergen's *Drosophila* group, for example, insisted that for "lower" organisms, such as fruit flies, an "artificial" environment had little or no effect on their capacity to produce natural behavior. Organismal complexity seemed to determine the importance of the "natural" experimental environment and methodologies; "the difficulty of supplying the requisite conditions for an animal to respond naturally in captivity apparently increases as one climbs up the phylogenetic tree."[46] Lorenz had also minimized the importance of field work, arguing that keeping animals in semi-captivity allowed him to observe their full behavioral repertoire and to identify more easily the functions of those behaviors—semi-captivity provided a perfect combination of freedom for the animals under observation and convenience for the researcher.[47]

In the early 1950s, British zoologists with interests in animal behavior saw themselves as participating in an enterprise that differed sharply from experimental psychology in the United States. After Danny Lehrman's critique of ethological methodology in 1953, Tinbergen—the figurehead of British ethology in the postwar era—also distanced himself from what he characterized as the rigidly hereditarian approach to innate behavior espoused by Lorenz.[48] The later 1950s and 1960s brought British ethologists and American psychologists into closer scientific contact. In the popular press, however, many of the theoretical assumptions of early ethology (already challenged by practicing ethologists) continued to be propagated with regard to both animal and human behavior.

Bower Birds: Ethological and Non-ethological Theories

Even before Tinbergen arrived in England, British zoologists such as Julian Huxley, William Homan Thorpe, and David Lack were keenly interested in promoting the study of animal behavior as an academic discipline within the biological sciences. Founder of the Animal Behavior subdepartment at the University of Cambridge, Thorpe later commented that, "from the zoologist's point of view, the coming of ethology was a reaction against the uncritical anecdotalism of much of the writing on animal behavior of the last century. With its emphasis on objectivity, ethology represents the scientific rebirth of one major part of natural history."[49] Huxley was no longer engaged in active research by 1950, but he maintained his connections with biologists interested in evolution and behavior. One of the men in whom he took a long-standing interest was an Australian, Alan John "Jock" Marshall, who arrived in September 1946, at the age of thirty-five, to read for a doctorate in philosophy in the Department of Zoology at Oxford University.

Eventually earning both his D.Phil. and D.Sc. at Oxford, Marshall centered the bulk of his research on the reproductive behavior and physiology of bower birds. For early students of animal behavior, male bower birds were intelligent aesthetes—driven by their artistic nature to colorfully decorate their elaborate mating arenas with flowers, shells, bits of fabric, or whatever they could find.[50] For zoologists seeking to distance themselves from amateurish speculation, aesthetically based explanations were simply unacceptable—and this extended to female choice. Marshall's research provided an evolutionary explanation for the bower birds' behavior that avoided naive anthropomorphism by framing their behavior as an adaptation to local environmental conditions.

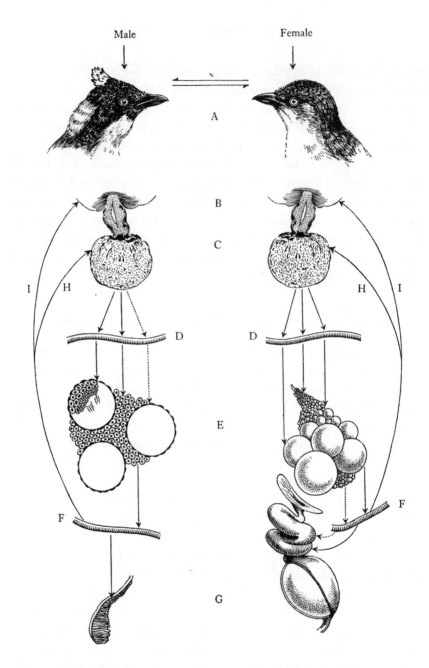

Male | Female

A

B

C

I   H

D   D

E

F

F

G

Jock Marshall suggested that the sexual behavior of the bower bird ("combat, sexual display, the discovery and acquisition of a mate, nest-building, and finally coition and actual reproduction") is a response to stimuli from the environment (*A*) that release chemotransmitters, which in turn act on the hypothalamus in the lower brain (*B*). The hypothalamus, in conjunction with the pituitary gland (*C*), directs the development of the sexual organs in both male and female birds (*D*, *E*, *F*, *G*). In other words,

In no way could Jock Marshall be called a traditional or average student. At the age of fifteen, he shot off his left arm while rustling some logs with the stock end of his shotgun. Much later in life, he used to quip that "it isn't necessary to have two arms—the weight and space would be far better taken up by an extra penis!"[51] Starting in 1929, at the age of eighteen, and unable to find work because of Australia's Depression, he spent four years jumping the "rattlers" (freight trains)—in other words, hitching rides on trains by grabbing hold of the moving car with his right arm and swinging the rest of his body up. In 1933, ensconced once again in Sydney and searching actively for employment through a connection at the Australian Museum, Marshall convinced Oxford professor John Baker that he, Jock, was the perfect man to travel to New Hebrides (Vanuatu) and assist with Baker's research on the highly valued intersex pigs bred locally. The different types of genitalia exhibiting both male and female characteristics were known by fantastically evocative names: "sprouting coconut, sewn-up, fruit bat, rat, hidden and worm."[52] Marshall's experiences working in the New Hebrides, collecting for the Sydney Museum, writing for the Oxford Exploration Club in New Guinea, and spending a couple of years in the United Kingdom lecturing to raise money for other expeditions, convinced him to enter the world of academic science. After getting married, earning his undergraduate degree at Sydney University, and serving as an intelligence officer in the Australian Imperial Force during the Second World War (earning him more experience in New Guinea under rather different circumstances), he arrived in Oxford, already divorced and eager to begin his studies.[53]

Bower bird behavior remained one of the last bastions of aestheticism in the animal kingdom. Male bower birds build a mating arena out of twigs, called a bower, which they then decorate with colorful objects—each species preferring a different color. Satin-bower birds, for example, prefer royal blue objects and will use anything from flowers to the feathers of other birds that they can find in the forest. A popular article written in 1944 described the underlying function of the male's behavior as having "gone far beyond" a "purely utilitarian usage." After watching the males building and decorating their bowers, "one cannot doubt that they derive

---

both the seasonal ripening of the sexual organs and the bower bird's complex courtship behavior could be explained as a mechanical reaction to environmental factors. The sex hormones (estrogen in females and testosterone in males) also feed back to regulate further sexual development (as controlled by the pituitary, I) and to elicit the production of innate behavior patterns by the brain (H). From Alan John Marshall, *Bower Birds: Their Displays and Breeding Cycles, a Preliminary Statement* (Oxford: Clarendon Press, 1954), 8–9, by permission of Oxford University Press.

a great deal of satisfaction and pleasure in such activities."[54] This aspect of play and aesthetic satisfaction set bower birds apart as unique, specialized, and exotic (their habitat distribution was limited to the southwest Pacific).[55] However, for zoologists interested in animal behavior, play was not an acceptable evolutionary cause.

Marshall published his treatise in 1954, *Bower Birds: Their Displays and Breeding Cycles, a Preliminary Statement*, and drew much of his conceptual framework from the professional community in which he circulated.[56] He shared with other Oxford zoologists a common understanding of mating display as a signaling device, a necessary precursor to successful mating. Marshall's official ties were to the reproductive sciences through his namesake and external examiner F. H. A. Marshall at Cambridge (who died in 1949, just as Jock finished his D.Phil.) and cytologist John Baker, his supervisor.[57] In a BBC radio broadcast in 1947, Jock mentioned F. H. A. Marshall's suggestion that "display is mutually stimulating between the male and female and that all the dancing and posturing and attitudinizing that seems so odd and striking to us, has the function of bringing each bird up to an equal physiological pitch so that when they mate the eggs are always fertile and a family results."[58] When Tinbergen arrived at Oxford in 1949, he and Jock Marshall also began to discuss bower bird behavior.[59] From Tinbergen, Marshall almost certainly took the idea that the birds' elaborate courtship bowers were derived from a displacement activity causing the birds to exaggerate their nest-building behavior.[60]

Julian Huxley became a particularly influential friend and mentor to Marshall, and they discussed the younger man's research extensively. Although Huxley and Marshall had met earlier, during Marshall's lecture tour of the United Kingdom in the 1930s, their friendship was cemented when Marshall led the 1947 Oxford University Expedition to Jan Mayen Island off the coast of Norway and Julian's son (Francis) went along.[61] In the 1950s, as Marshall was working on his book, he lived close to the Huxleys and they shared many discussions about the bower birds' behavior.[62] Marshall's views on sexual selection were strongly affected by these conversations. He agreed with Huxley that male display functioned to increase female sexual excitement and should not count as aesthetic sexual selection, because a minimum level of sexual interest was a necessary precursor for a successful mating.[63] Without the necessary ritualistic behavior, the proper union of gametes would not occur and the individuals would not reproduce. Huxley, let us remember, contended that not only were most attractive male traits due to natural selection rather than sexual selection, but in the few cases where sexual selection did play a role, it was through the action of male-male competition, not choice-based female preferences.

Marshall, for his part, was fascinated by the possible correlations between the bower birds' exotic behavior and their internal reproductive physiology. In his book,

Marshall suggested that although "a form of sexual selection probably occurs," the displays of the bower birds increased their chances of surviving and reproducing—it synchronized the birds' sexual cycles in an ecologically variable environment.[64] The courtship and display of the male bower bird cemented the pair bond and kept the female at the nest throughout the courtship period. The construction of the bower itself he attributed to a displacement activity while the male waited weeks for the female to come into sexual readiness. Marshall framed the exaggerated courtship behaviors as serving an evolutionary function previously unrecognized by zoological investigators.

Marshall's adept dismissal of recreation and conscious aestheticism as the primary causes of bower-decorating behavior generated his peers' respect for the book. *Bower Birds* was well received precisely because he provided a new way of conceptualizing the mating behaviors of these unusual birds as the result of ecological and behavioral rituals, rather than desire. Almost all reviews of Marshall's book contrasted his scientific approach to animal behavior with older anthropomorphic descriptions of bower bird behavior. A review in the London *Daily Telegraph and Morning Post* applauded him for rejecting amateur notions of aesthetic play in bower birds, saying (as a compliment) that "on these wonderful materials for poetic writing Dr. A. J. Marshall pours a cold douche."[65] Later reviews by ethologists expressed similar sentiments about Marshall's display-not-play theory. One suggested that, after Marshall's book, "'intelligence' and conscious application in bower-birds must be rejected together with much nonsense that has been popularised about them. They present an extravagant example of the amazing complexity of behaviour which instinctive pattern can initiate and control."[66] Another review contrasted Marshall's sober scientific analysis with previous "anthropomorphic nonsense."[67] In the minds of his fellow biologists, Marshall had successfully substituted instinct and sexual necessity for intelligence and poetry in explanations of bower bird behavior. Marshall's reliance on natural selection and male-male competition rather than female choice allowed him to steer clear of discussing potentially anthropomorphic behaviors.

Marshall's explanation of bower bird behavior also emphasized the evolutionary causes of courtship behavior. By recasting courtship as mutual stimulation required for proper procreation, he circumvented discussions of how female choice could act to alter the male's behavior or appearance. Although all subsequent investigations of bower-building behavior adhered to Marshall's display-not-play formulation, female choice would eventually return as a viable explanation. Some authors, including E. Thomas Gilliard at the AMNH, suggested that bower structure was a physical extension of the male bird, designed to attract females. Gillliard noted that physical differences between the sexes were less pronounced in species with more complex bowers than in species with simpler constructions: "I believe that the forces of sexual selection in these birds have been transferred from morphological characteristics—the male plumage—to external objects and that this 'transferral effect' . . . may be the key factor in the evolution of the more complex bower birds."[68] Gilliard endorsed female choice as an evolutionary mechanism, with no concern about the potentially anthropomorphic connotations of "choice" in nonhuman animals.

Even though Marshall was not an ethologist, he drew his ideas about the evolution of behavior from the ethological community at Oxford. Thus, natural selection controlled behavior in his theories of bower bird evolution, leaving no space for female choice in which the courtship behavior of individuals could feed back and affect the evolutionary process itself.

Fruit Flies: Ritualized Courtship in a Model Insect

Another research organism that proved important for studying reproductive behavior in the decades following the Second World War was the fruit fly, *Drosophila*, and this research took place under the nominal guidance of Tinbergen at Oxford. As Tinbergen preferred to be out in the field rather than trapped inside a laboratory, his graduate students working on *Drosophila* labored more or less independently, consulting a wide range of zoologists and geneticists for guidance in their work.[69] They conducted their research in a laboratory, obtained their organisms from standardized breeding stocks, viewed their subjects through microscopes, and at times performed invasive surgery, all in the name of understanding the flies' "natural" behavior. Tinbergen's students, like many other zoologists at the time, considered fruit fly behavior natural under these conditions because of the simplicity of the insect brain. Flies, they argued, could not perceive that they were in an artificial environment and thus would behave similarly under a microscope and outside, among the leaves of a tree. For geneticists, fruit flies may have been a standardized laboratory tool, a kind of living black box,[70] but for Margaret Bastock and Aubrey Manning, fruit flies were just another kind of natural organism. Perhaps because zoologists conducting research on fruit fly behavior were less concerned with distancing themselves from anthropomorphism than were their colleagues working on other animals, drosophilists investigated questions of female mating preferences and the physical and physiological bases of those preferences more freely than did zoologists, like Marshall, who worked on more charismatic fauna such as bower birds.

The initial inspiration for an ethological analysis of fruit flies' mating behavior came from a perhaps unlikely source—ornithologist Ernst Mayr. Mayr now occupies the pantheon of great twentieth-century evolutionary biologists, but in 1949, he was a moderately well-known zoologist working at the American Museum of Natural History who had, in recent summers, taken to dabbling in *Drosophila* research.[71] That year, Mayr wrote to ethologist Konrad Lorenz: "It may amuse you to learn that I too have recently worked in the field of animal psychology. However, I have worked with little fruit flies, *Drosophila*, rather than with birds. Actually, there is not as much difference as you might imagine."[72] An avid correspondent, Mayr also sent a series of letters to Tinbergen extolling the virtues of fruit flies as ideal organisms in which to study the dynamics of species recognition. In fact, the only disadvantage Mayr mentioned in his letters was that the flies were, "to be sure, . . . quite small, but this disadvantage is far outweighed by the fact that they can be studied under the binocular microscope."[73]

Mayr conducted some of the earliest experiments on determining the physical basis of female mate recognition in *Drosophila*.[74] He surmised that female choice in flies might be based on olfactory cues. The antennae of fruit flies contain their scent organs, and Mayr surgically removed the antennae from female flies to see whether, without a sense of smell, they would retain the ability to differentiate among individual males. They did not. In fact, when Mayr placed the antenna-less females in vials with males of two species of fruit fly (*D. pseudoobscura* and *D. persimilis*), they mated with each kind of male in equal proportion.[75] Mayr originally hypothesized that if antenna-less females could not distinguish between males of different species by smell, then his operation should increase female acceptance of males. However, he found exactly the opposite: antenna-less females were *less* likely to mate than were normal females. When Mayr found himself unable to continue his summer work on *Drosophila* at Cold Spring Harbor, he suggested to Tinbergen that someone else should continue to investigate mating behavior in fruit flies. "I certainly wish," he pleaded, "that some qualified person, say a student of yours, would become interested in this work and carry on."[76]

Within a year, Tinbergen had a graduate student working diligently on the mating behavior of fruit flies—Margaret Bastock.[77] Aubrey Manning soon joined Bastock in her efforts. Tinbergen eventually wrote to Mayr that "your repeated prodding about *Drosophila* has had some effect!"[78] By 1960, Bastock and Manning were married and had moved to Edinburgh. At Edinburgh University, they established a new center for the ethological investigation of the genetic basis for behavior, and their move signaled the end of most *Drosophila* research at Oxford under Tinbergen's direction. Tinbergen himself was never particularly interested in the mechanics of the *Drosophila* research—he was much more interested in being outside.[79] When Bastock and Manning left for Edinburgh, they took with them their tacit knowledge of how to maintain *Drosophila* stocks and observe the flies' mating behavior. For Tinbergen, their move probably made training new graduate students in *Drosophila* research at Oxford extremely difficult. Any potential graduate student interested in ethology and *Drosophila* behavior would have enrolled at Edinburgh to work with Manning.

While Bastock and Manning were at Oxford, the mating behavior of fruit flies proved to be an extremely productive line of research. Manning provided two main reasons that *Drosophila* was so useful as an ethological research organism. First, he suggested, ethologists could easily separate the "limited effects of learning from the rich endowment of instinctive behaviour in insects," making fruit flies perfect subjects for studying the evolution of behavior. Second, fruit flies exhibited incredibly elaborate courtship displays and exemplified "all the major features of behavioural evolution."[80] Ethologists' acceptance of fruit flies as organisms worthy of their attention is also borne out in the frequency of *Drosophila* papers in the ethological journal *Animal Behaviour*—between 1958 and 1975, it published thirty-three papers on the behavior of fruit flies.[81] Despite our common association of *Drosophila* with standardized laboratory research, fruit flies were certainly not unusual subjects for ethologists.

Fruit flies proved most useful to Tinbergen's research group in answering Danny Lehrman's 1953 critique of the instinctual basis of behavior. In her 1956 paper, "A Gene Mutation Which Changes a Behavior Pattern," Bastock suggested that although ethologists believed behavioral traits evolved in the same way as physical traits, until someone demonstrated a genetic basis for behavioral variations, that belief was simply an assumption.[82] Bastock showed that the genetic mutation causing a "yellow" phenotype in some *Drosophila* stocks correlated with differences in mating behavior—specifically, the frequency and duration of wing vibrations during the courtship dance. Perhaps, Bastock argued, genetically based behavioral differences such as these were a first step in developing reproductive isolation between two populations.[83] Bastock's paper gave ethologists an experimental basis on which to claim that behavioral traits in other organisms were also heritable and therefore could be affected by natural selection. Several zoologists cited Bastock's paper as evidence that genetic differences probably underlay many behavior patterns, and her paper was most heartily received by the burgeoning community of behavioral genetics (to which we turn in chapter 5).[84]

Before he arrived at Oxford, Manning had obtained his undergraduate degree at University College London, where he worked with geneticist J. B. S. Haldane and became good friends with another of Haldane's students, John Maynard Smith. When Manning left London to study under Tinbergen for his dissertation research, Maynard Smith continued working with Haldane on the courtship behavior of

Drosophila. Maynard Smith, Manning, and Bastock maintained their friendship, sharing ideas throughout their careers, although their interpretations of *Drosophila* courtship differed considerably.[85]

In his early research, Maynard Smith attributed the lack of reproductive success of some *Drosophila* males to a reduced athletic ability and corresponding inability to keep up in the mating dance: unsuccessful males simply could not provide enough stimulus to the female to induce her to mate—they tried their best and failed. It was an issue not of male motivation, but lack of capacity. He later quipped that it was a good thing that human courtship did not necessarily work the same way, or he might have remained single![86] Additionally, some male flies would not be able to secure a mate, because females preferred signals that were more extravagant than necessary. In his analysis of the courtship dance of *D. pseudoobscura,* Maynard Smith listed three types of reproductive advantage for females who were unwilling to mate with every male they encountered (he did not consider these advantages to be mutually exclusive): (1) pickiness ensured that females mated with a male member of their own species, (2) pickiness could increase the number of female partners with which an outstanding individual could mate, and (3) pickiness could help females find a single high-quality mate who would help them produce more offspring (compared with a lower-quality mate). These advantages—functions, really—of choosy females implied not only that a female would benefit from the ability to distinguish males of her own species from males of other species, but that she would also benefit from discerning minute differences in quality among males of her own species.[87]

When Manning followed up on Mayr's preliminary investigations, however, he couched the courtship behavior of male fruit flies as inherently stimulatory in nature and ignored Maynard Smith's suggestions about the function of mate choice in flies. True, Manning felt that the question at hand was how (on what physical basis) and why (theoretically) some male flies were more successful than others at acquiring female mates. However, he placed the onus for these discrepancies on the female's perception of male stimulation—females were not being choosy, there was simply a breakdown of courtship signaling. By searching for direct physical mechanisms of fruit flies' mating behavior, Manning tried to uncover how stimuli were transferred from one fly (male) to another (female), thereby causing a behavioral response (copulation). Zoologists believed that most fly-to-fly communication occurred through scent or hearing (by "hearing," ethologists meant the ability of flies to perceive vibrations in the air). Both of these senses rest in the antennae, and Manning (following Mayr's precedent) removed pieces of the flies' antennae to see more specifically which parts were responsible for communicating the movements and

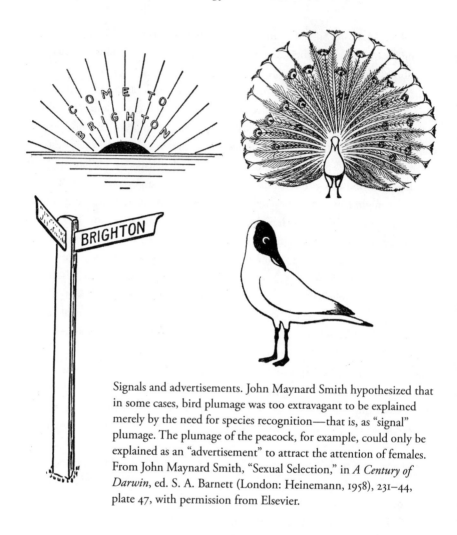

Signals and advertisements. John Maynard Smith hypothesized that
in some cases, bird plumage was too extravagant to be explained
merely by the need for species recognition—that is, as "signal"
plumage. The plumage of the peacock, for example, could only be
explained as an "advertisement" to attract the attention of females.
From John Maynard Smith, "Sexual Selection," in *A Century of
Darwin*, ed. S. A. Barnett (London: Heinemann, 1958), 231–44,
plate 47, with permission from Elsevier.

smells of one individual to another. Despite their invasive nature, Manning valued
these experiments as elucidations of normal mating behavior, as long as they were
combined with direct observation and analysis of the resulting courtship behavior.

Biologists trained as population geneticists used different methods of observation than did the ethologically trained Bastock and Manning. In 1958, the French
geneticist Claudine Petit also followed up on Mayr's experiments, trying to isolate
which parts of the antennae were responsible for female choice. She performed four
different operations on female *D. melanogaster*, in each removing a different portion of the antennae. Like Mayr, Petit placed her postoperative females in vials with
males for a long time (four days; Mayr left his for two days) and then checked to

see which females had been inseminated by the end of that period. Manning found this approach appalling and was particularly astounded that both Mayr and Petit had merely calculated the percentage of "successful" matings without watching the courtship itself. This was not an appropriate technique for investigating behavior, he argued, because the males would become more and more desperate the longer they were confined in a vial with a female, and they would eventually resort to unnatural forced copulations, thereby skewing the results.[88] Manning's critique did not hold much water with Petit, who remained convinced that forced copulations did not occur in *Drosophila*, even in a laboratory. She argued that a female *D. melanogaster* had the ability to reject a male decisively by everting her genital duct with a loud "pop."[89] Males presented with this behavior, she said, stopped courting immediately. Despite Petit's belief that Manning was wrong about *why* his technique was better, she traveled to Oxford to learn the recording techniques for direct behavioral observation developed by Bastock and reported by Manning in his paper.

Meanwhile, Manning repeated all of Petit's published experiments. He watched pairs of mating flies for fifteen minutes each, and if the male courted and copulated with the female during that time, he scored the mating as "successful."[90] Manning discovered that two parts of the antenna—the funiculus and the sail-like arista—acted as a unit, twisting in response to vibrations in the air and stimulating the Johnston's organ (in effect, the ear of fruit flies). Whereas Mayr suggested that flies reacted to the pheromonal *smells* of other flies, Manning concluded that the beating of male fruit flies' wings during courtship acted as a *physical* stimulus to the female, increasing her sexual excitement. Manning's mechanical model of *Drosophila* courtship differed from Mayr's only in the kind of stimulus required to cause courtship behavior in the female (auditory, rather than olfactory). The naturally reticent females needed stimulation to mate, without that stimulation (without their antennae), their cooperation plummeted. By providing a model of mating behavior in flies that avoided Maynard Smith's choosy females, Manning and Bastock's model of *Drosophila* courtship further reflected their ethological training.

Based in part on Bastock and Manning's research, Tinbergen noted in his class lectures that *Drosophila* courtship served six distinct ritualized functions.[91] It allowed for the proper orientation of male and female genitalia during copulation. It ensured that males and females were ready to mate at the same time—it synchronized their desire. Ritualized courtship behaviors could act to persuade both the male and female to mate rather than attack each other or run away in fear. Species-specific mating behavior ensured that only members of the same species would be sufficiently stimulated to mate. Courtship behaviors often took the form of pre-parental care—building a nest, for example. And, finally, mating behavior served to reinforce pair bonds between mates; this was particularly important for mating displays that continued after copulation had occurred.[92] In each of these functions, behavior was acted on by *natural* selection. For Bastock, Manning, and Tinbergen, the function of a male's courtship behavior was not, in the sense of sexual selection, to convince females to mate with him, but, in the sense of natural selection, to ensure the proper proliferation of the species.

From 1950 to 1965, the major issue for the ethologically inclined zoologist remained the study of how evolution had acted to modify observable behavior in animals. Fruit flies, in part because of their past use in genetics research, were easily adapted for ethological research. Bastock used flies to demonstrate the genetic basis of courtship behavior, an analysis that was practicably impossible in any other species. Manning demonstrated the importance of stimulation in coaxing a female to mate. So, on the one hand, zoologists also appreciated the unique nature of fruit fly courtship due to the importance of nonvisual cues and, they believed, its completely instinctual basis. On the other hand, fruit flies functioned as a model stand-in for other species in which the genetic basis of behavior could not be worked out.

Ritualized Human Courtship

The translation of Lorenz's popular books into English, including *King Solomon's Ring* in 1952 and *On Aggression* in 1966, inspired an entire genre of "pop ethology" in which zoologists applied their disparate ethological frameworks to human behavior.[93] All of these books were controversial, and any number of scientists and popular readers criticized pop ethologists for their easy slippage between animals and people.[94] Tinbergen insisted that ethologists should take the high road by applying their *methods* of behavioral analysis to humans, rather than (like psychologists) simply generalizing the *results* of observations of animal behavior directly to humans.[95] Yet even this zoological approach, adopted explicitly by Desmond Morris in *The Naked Ape* (1967), proved incendiary. Social scientists had been honing their professional identity, analytical techniques, and theories of observing human social behavior for decades and did not look kindly upon such interlopers.[96]

After the Second World War, zoologists vigorously pursued the question of what makes us human. The decolonization of Africa and South Asia allowed greater access to natural field sites for growing numbers of zoologists interested in both human and primate behavior.[97] For zoologists interested in the evolution of human social behavior, no question was more pressing than human aggression. When issues of sexual relations and courtship behavior did arise, they were often linked with questions of territoriality, conflict, and the seemingly innate and unique tendency of humans to make war and wreak violence on their own species.[98] These zoologists sought to turn their knowledge about animals into a source of information on human behavior without invoking a racial hierarchy, by searching for stereotyped behaviors shared by all human (and sometimes ape) societies: "territoriality, optimum population maintenance, agonistic behaviour, dominance and hierarchy, bonding, epimeletic behaviour, mating and consort behaviour, ritualised display, play, intergroup relations, communication systems, etc."[99] By concentrating on human behaviors categorically, zoologists could analyze these behaviors as adaptations—as heritable units that changed over time based on the environmental circumstances of the species—and hypothesize about the role of those behaviors in the survival of the individual and/or the group in which they were expressed.

In this sense, zoologists found it easy to analyze humans as simply another animal, uniquely adapted to its natural environment, exhibiting the same categories of behavior as other species: foraging behavior, courtship behavior, territoriality. Yet humans also represented a methodological problem in terms of identifying normal human behavior under natural conditions. Self-defined human ethologist Irenäus Eibl-Eibesfeldt looked for human nature primarily in cultures "untouched" by Western civilization, the members of which, he presumed, were living closer to the natural conditions characteristic of early hominids.[100] Yet Eibl-Eibesfeldt added another condition as well: people of any culture behaved naturally only when they did not know they were being watched![101] He used a special sideways lens on his movie camera that allowed him to film people's behavior on the sly; his behavioral subjects thought he was recording the action in front of him, not off to the side. Morris's zoological analyses of human behavior, however, focused almost exclusively on people from his own culture—London, England. Lorenz, by way of contrast, talked with equanimity about both "civilized man" and "Cro-Magnon warriors."[102]

The concept of ritualization again proved useful and enabled ethologists to search for similarities in behavioral patterns among large numbers of species, but ritualization became problematic when ethologists used it to describe social interactions in people.[103] In 1966, Julian Huxley had cited two functions of ritualization in animals—communication and bonding.[104] For the purpose of communication, he suggested, ritualization reduced complex information about an individual's emotional state to an almost automatic reaction, a behavioral reflex, that the recipient of such a signal could instantaneously interpret. Because behaviors acted to bond two individuals over time, ritualized signals became more intricate and temporally extended. Ethologists applied both kinds of ritualization to human behavior, often (but not always) linking communication to territorial aggression and bonding to courtship.

Many zoological treatises on human behavior took as their central theme the human capacity for individual and coordinated aggressive behavior, a problem of territorial instincts gone awry. Lorenz argued in *On Aggression* that modern weapons killed people at a distance and with such ruthless efficiency that they circumvented the victims' opportunities to proffer ritualized signals of submission; technological advance had outstripped human behavioral evolution.[105] As a result, humans, like doves, lacked the common sense not to peck each other to death when crowded into a cage.[106] In captivating prose, Lorenz wrote that "with atomic bombs in its hands and with the endogenous aggressive drives of an irascible ape in its central nervous system, modern humanity is a whole or system which has got very thoroughly out of balance."[107] Tinbergen, for his part, agreed with Lorenz's basic point that cultural evolution was outpacing the physical evolution of human society and that humans needed to overcome their innate nature in order to live in an orderly and productive society.[108] Tinbergen remained more hopeful than Lorenz, insisting that humans were different from other animals because they exhibited an extraordinary ability to change their environmental conditions and thus the selective pressures under which they lived—through science, humans could understand and learn to control their aggressive drive.[109] Other ethologists were less sympathetic to Lorenz's enterprise. Robert Hinde, for example, argued that in writing *On Aggression*, Lorenz ignored all the criticisms developed by ethologists about the study of "instinct" in the preceding decade (especially those articulated by Danny Lehrman in 1953): behaviors could not be categorized as either "innate" or "learned," aggression was not spontaneously produced in individuals as an instinctual drive, and "aggression" in animals and humans constituted different phenomena. Another book often mentioned in the same breath as *On Aggression* was written by playwright Robert Ardrey and described the research of paleoanthropologist Raymond Dart: *African Genesis*. In the tradition of Freud, Ardrey argued that humanity came into being when we learned to kill (an image memorialized by Stanley Kubrick's film *2001: A Space Odyssey* in 1968). An especially perceptive reviewer of *African Genesis* noted that all the hubbub over territorial aggression and defense as basic to human identity seemed to be "a condemnation, as against nature, of Communist society."[110]

Yet love (in the form of courtship) proved just as important as aggression to Irenäus Eibl-Eibesfeldt's picture of human nature. Eibl-Eibesfeldt emphasized the importance of communication in driving the evolution of courtship rituals in birds and humans. He began by arguing that the extravagant tail coloration of the male peacock evolved as a ritualized form of "food enticing behavior." As evidence, he cited, among other things, the male's behavior of scratching on the ground in front of the female to attract her attention, and the behavior of the female peacock who

Irenäus Eibl-Eibesfeldt suggested that when humans flirt, they first meet the eye of the object of their affections (or, in this case, the camera) and then quickly glance away, inviting pursuit. These images are stills from his ethnographic films. From Irenäus Eibl-Eibesfeldt, *Love and Hate: The Natural History of Behavior Patterns*, trans. Geoffrey Strachan (New York: Aldine de Gruyter, 1970), figs. 17 and 19.

then came to search for food, only to find herself "at the focal point of the 'concave mirror' formed by his tail-fan." For Eibl-Eibesfeldt, the effect of this ritualized courtship was to signal the emotional state of one animal to another—to indicate their interest in mating. Similarly, he suggested that as human behaviors were codified into "rituals," they also became both more simplified and more exaggerated "in the manner of a mime."[111] He was especially fascinated by the way women flirted with their eyes. He described how, in many cultures, a woman would gaze beguilingly at the object of her affections and, if their eyes should perchance meet, would quickly look away, symbolically fleeing the scene and inviting the observer to pursue. The human behaviors of interest to Eibl-Eibesfeldt were determined by his ability to generalize from animal to human, and this limited the range of human behaviors he chose to investigate to ritualized, automatic responses. Eibl-Eibesfeldt's friend and fellow human ethologist Hans Hass noted similar differences between male and female roles in human courtship: "In men, a mood of aggression intensifies sexual appetency, whereas fear diminishes it. In women, aggression diminishes sexual preparedness, whereas fear can intensify it."[112] Hass explicitly linked male aggression with sex drive—male pursuit and female retreat. Mate "choice" never entered the picture, even with humans.[113]

Desmond Morris, a former graduate student of Niko Tinbergen's, published several best-selling books on human behavior, including *The Naked Ape*, in which he also emphasized the importance of sex in defining humanity.[114] In *Naked Ape*, Morris intended to analyze the human species as would a Martian just recently arrived on Earth. The book contained lengthy chapters on sexual behavior, and Morris suggested that human sexual partnerships were the source of modern social cooperation: ritualized bonding began at the level of the family, expanding over evolutionary time to include nonrelatives. Based on observations of British citizens, Morris described humans as the "sexiest" apes, positing that, as a species, we engage in pleasurable sexual interactions far more often than the other ape species.[115] Because they engaged in sexual play more frequently, Morris suggested, humans were more altruistic than other ape species. He maintained that social groups of humans exhibiting monogamy were more likely to suppress aggression within the group and therefore to be more successful in turning those attentions to other, productive enterprises, such as building an economy, science, and so on. Additionally, groups with less intragroup aggression, he postulated, would be more likely to put on a unified front when confronted with an external threat and to win intergroup aggressive encounters. For Morris, the key to our biological future was to make love, not war. One cranky reviewer commented that his "only serious objection to that delightful book was that it left out almost everything—language, abstract reasoning, art, institutions, etc.—that distinguished man from other primates, rats, ants, worms, asparagus."[116] This remark captured most of the major criticisms of Morris's approach—namely, that he seemed to imply that humans were "nothing but" naked apes.[117]

Not all popular takes on the human species held the same rosy view of human sexual relations. Anthropologically trained Robin Fox, for example, knew Morris well, as they both held posts at the London Zoo studying animal behavior.[118] Fox insisted that male-male competition, not sexual relations, formed the basis of modern human social structures by defining the social dominance hierarchy in which all individuals fit. Irreverent at heart, and at times deliberately provocative, Fox insisted that Morris's correlation of "pair bonds" in animals with "falling in love" in humans was overly optimistic. Fox preferred to think of human marriage as a contractual, economic affair. The reason people stayed married, he suggested, was more due to their role as parents than as lovers: "Love and marriage may go together like a horse and carriage, but let us not forget that the horse has to be broken and harnessed."[119] The fiscal, as well as moral, messages were clear in Fox's historical reconstruction of the evolution of human social relationships.

As a model for understanding human social interactions, Fox contended that baboon troops were regulated through a strict dominance structure—only the top males got to mate with females during estrus, and therefore only the top males passed along their genes to the next generation. Why did some males succeed where others did not? Fox suggested "smartness." Males who were smart enough to delay sexual and social gratification would eventually succeed, while those who were driven to mate with the female rather than temporarily defer to the dominant male were doomed to become the "drop-outs of the monkey rat race." Fox continued, "the stupid animal . . . that blunders about, following without foresight the dictates of his lustful and aggressive appetites, will never make it to the top."[120] In Fox's argument, the ability of *males* to negotiate the dual desires for sex (from available females) and social status (through aggressive competition with other males) had driven the intellectual evolution of humans and other primates. What defined us as human—our ability to reason, to negotiate, to operate in a highly complex and constantly shifting social milieu—arose from these male animalistic instincts. Women contributed to the evolution of intelligence by making themselves constantly available for sex, thereby forcing men to provision them with food brought back from the hunt and thus establishing semi-permanent family relationships.[121]

In the 1960s and 1970s, pop ethologists continued to equate animals with emotion and, conversely, rationality with humanity in ways that are reminiscent of Fisher's and Huxley's earlier desires to maintain a bright line between animal and human, male and female. However, whereas Huxley and Fisher argued that social structure in humans arose through female choice (because only in humans could individuals truly "choose" their mates), much of the postwar zoological literature on the evolution of human social systems suggested that the social structure of human civilization derived instead from male-male competition.[122] Because of this emphasis on looking for explicit connections between the ritualized behaviors of animals and humans, choice-based behaviors never entered their explanatory framework.

British zoologists' descriptions of human courtship drew strongly on elements of behavioral analysis developed to discuss animal behavior. In many cases, these elements harkened back to a mode of "classical" ethological analysis already under criticism within the zoological community: Lorenz's assertion that an instinctual behavior was something that animals either "have got" or are lacking; the belief that instinctual behaviors were categorically separate from "learned" behaviors; the belief that a hydro-mechanical model of behavior could explain why the drive to act could build up until it must be released, if not by an innate releasing mechanism, then by an unrelated displacement behavior. When applying these to models of human courtship, zoologists thereby emphasized the importance of courtship for pair bonding and the ritualized aspects of courtship. In the end, this meant that British zoologists after the Second World War, although fascinated by courtship behaviors, never emphasized the role of choice in their theoretical models of behavior production in either animals or humans.

The human-oriented books of the so-called pop ethologists were successful and wildly controversial. Despite the controversies over the theoretical basis of their claims about human behavior, they had begun opening a disciplinary space for zoologists once again to talk about the behavior and sociality of humans. In 1973, Niko Tinbergen, Konrad Lorenz, and Karl von Frisch were awarded the Nobel Prize in Physiology or Medicine. This mark of legitimacy attested to zoologists' success in claiming the right to talk about human behavior. In the presentation speech, the audience was told that "the discoveries made by this year's Nobel prize laureates were based on studies of insects, fishes and birds and might thus seem to be of only minor importance for human physiology or medicine." However, the presenter continued, the importance of the psychosocial environment and an animal's biological equipment in producing behavior "holds true for all species, also for that which in shameless vanity has baptized itself 'Homo sapiens.'"[123]

In their research on animal behavior, zoologists did not consider the concepts of "artificial" and "natural" to be coextensive with "laboratory" and "field." *Drosophila* behavior in a laboratory setting was natural. Bower bird behavior in the Australian bush was natural. Unobserved human behavior in London was natural. The flexibility of "natural" as a category allowed zoologists to combine research on different organisms, conducted with a variety of research methods, into a multifaceted vision of how evolution could act to modify animal behavior over time.

With the bitter debates over organismal and molecular approaches to research starting in the late 1960s, field research became increasingly important in the study of natural processes. As *Drosophila* behavior is almost impossible to observe in the field, fruit flies were branded artificial laboratory creatures. The role of *Drosophila* in the history of ethology was substantially discarded as a result of their new, non-natural, status, while the history of ethology in birds and fishes was increasingly emphasized. Research on bower birds also benefited significantly from the discovery that polygamy was far more common in natural populations than was previously thought.[124] The fluid boundary between laboratory and field investigations that characterized earlier ethological research became more difficult to maintain. Thus, it is to the transformations wrought by the laboratory context of behavioral genetics that we next turn.

CHAPTER 5

A Science of Rare Males

The Genetics of Populations in the Long 1960s

One aspect of sexual selection, in the original Darwinian sense, is the evolution of secondary sexual characters. Another even more important aspect is the reproductive isolation between species by sexual selection. From an evolutionary point of view, sexual isolation is the chief function of sexual selection, that is to say, of sexual discrimination. There is a third important function of sexual selection, viz. the maintenance of genetic heterogeneity in populations as a consequence of the advantage of heterozygotes in sexual selection.

—Ernst Bösiger, "Role of Sexual Selection," 1974

In 1946, a young French scientist named Claudine Petit entered graduate school at the Centre National de la Recherche Scientifique (CNRS) in Paris. For her dissertation research, she decided to investigate *selection sexuelle* in the fruit fly *Drosophila melanogaster*. She published a series of articles throughout the 1950s, including her description of a "rare-male" mating advantage in *Drosophila*. Females, she argued, preferred to mate with exotic males; this tendency illustrated true female choice, because it could not be explained by variations in the vigor of male courtship.[1] This story is remarkable for any number of reasons. Petit articulated a vision of female choice in insects that assumed female flies possessed the cognitive apparatus necessary for choosing; Petit is a woman; and she is French.

Why was choice-based behavior in flies so surprising? Following the Second World War, the numbers of biologists studying populations of *Drosophila* had steadily increased. The population geneticists who studied *Drosophila* sought to understand how genetic differences spread in a population, rather than investigating the causes, effects, and heredity of mutations in individual development as had the "classical" geneticists trained by Thomas Hunt Morgan.[2] In the genetic analysis of populations, the small, black-bellied fruit flies acted as human equivalents, as stand-ins for you and me.[3] At the same time, population geneticists capitalized on the association between insect minds and instinctual mechanization. Although fruit flies were not social insects (as exemplified by the self-sacrifice of ants or bees for the good of the nest/hive), insects generally represented an evolutionary path to *instinctual* sociality that acted as a foil for the *cultural* basis of human sociality.[4] Flies seemed to lack the cerebral or intellectual capacity to learn, and so population geneticists came to depend on flies for their ability to behave naturally in the most unnatural of conditions, including being raised by the hundreds in small glass vials filled with a gelatinized mixture of yeast, unsulfured molasses, and agar. Because of the obvious differences between insects and people, fruit flies offered a way for geneticists to investigate the process of evolution without overt eugenic overtones. Yet the similarities in the genetic structure of human and insect populations allowed these same biologists to generalize from their results on fly evolution to probable mechanisms of evolution in humans. Petit's description of "rare-male" mating preferences in female fruit flies did not sit easily within this framework in which fruit flies mated mechanically and humans chose their mates.

Petit's research was also unusual because she is French. Most biologists in France before the Second World War were reluctant to adopt Mendelian genetics.[5] Petit's advisor was a notable exception, and his work on genetic diversity of fruit fly populations formed one of the few French programs studying population genetics in the 1940s.[6] After completing her dissertation research, Petit formed connections with the population genetics community in the United States, especially with the help of Theodosius Dobzhansky. Her influence on the population genetics community, therefore, arose from her international ties to American geneticists rather than domestic ties to biologists in France.[7]

When Petit began her research, there were very few female population geneticists in charge of their own laboratories. Yet the number of women interested in sexual selection research in fruit flies at this time was larger than one might expect—two! Petit formed a long-lasting collaboration with another woman working in population genetics, Lee Ehrman. They met when Ehrman was a graduate student in Dobzhansky's laboratory and shared many research interests—among them, female choice. In interviews, Petit and Ehrman each insisted that their early interest in female choice had nothing to do with the fact that they were women; female choice was simply an open question that needed to be answered. Despite the difficulties facing women with research careers in the 1950s and 1960s, both Petit and Ehrman obtained university positions and led their own research laboratories. Petit suggested that her experience in the Resistance during the war gave her the courage to finish her degree and become one of the first female professors at the Université Paris VII. Ehrman became professor of biology at the newly founded Purchase campus of the State University of New York in 1972, and she has served as president of both the American Society of Naturalists and the Behavior Genetics Association.[8]

Petit's and Ehrman's research on the rare-male effect was unusual for its time, and their experiments sat at the intersection of several important research questions that characterized American population genetics in the long 1960s. Research on the role of behavior in evolution proliferated in the United States, although along remarkably different lines than had British zoology. Population geneticist Ernst Bösiger, for example, described several distinct dimensions to the function of reproductive behavior in evolution: reproductive isolation, the maintenance of genetic diversity in a population, and sexual selection—in "the original Darwinian sense."[9] Of these, he considered reproductive isolation to be the most interesting and sexual dimorphism the least important.

Experimental population geneticists with a theoretical bent found female mate choice fascinating for its possible role in the process of speciation, the splitting of a single interbreeding population into two reproductively isolated populations. Why? On the one hand, zoologists such as Ernst Mayr argued that all speciation must start with geographic isolation. On the other hand, theoretical biologists such as John Maynard Smith argued that if part of a population were to suddenly change its mating behavior, then the previously unified population might split into two independently breeding populations without any physical separation, a process known as "sympatric speciation." The question was, how might such a change occur? Changes in female choice represented one of the most likely mechanisms for sympatric speciation and provided an avenue for biologists to conceive of speciation without geographic isolation.

Yet by far the bulk of experimental population geneticists' interest in female choice was directed at understanding how mating preferences might act to change the genetic diversity of a single population. Petit and Ehrman argued that female choice of rare males could increase genetic diversity by preventing the extinction of the rare alleles of these exotic males in the breeding population. The larger the variety of breeding males in a population, they posited, the larger the genetic diversity of the population. In the context of evolutionary theory, genetic diversity is a good thing because it makes species more resilient to adverse environmental change. For example, if a population's environment were to change, or if a deadly disease were to spread through the population, a species with high genetic diversity would be more likely to contain individuals that were slightly more adapted to the new conditions or more resistant to the disease. Over time, these individuals would survive and produce more offspring, and the population as whole would adapt to the new environment or be more resistant to the disease. If a species with low genetic diversity were to encounter the same environmental change or disease, the species would be less likely to contain individuals able to adapt to change or resist disease and thus would be more likely to become extinct (think of the failure of the potato crop that caused the Irish potato famine). So, although the advantages of a genetically diverse population were clear, population genetics models illustrated that natural selection could not create or even necessarily maintain genetic diversity in a population. Strong natural selection should act to reduce the variety of breeding individuals in a population, so, over time, genetic diversity should always decrease. Female choice, through the rare-male effect, offered a possible solution. If females preferred to mate with males whose genetic attributes occurred infrequently in the population, then female choice of those males as mating partners would ensure that the males' rare genes were passed on to their offspring. For Petit and Ehrman, female choice acted to counter the effects of natural selection by preserving the genetic attributes of rare males for future generations.

The possibility that female choice could act *within* a single species differed radically from mid-century evolutionary models that emphasized the importance of choice solely in the context of reproductive isolation *between* species. In the late 1960s, mathematically and theoretically inclined population geneticists also began to investigate the possibility that female mating behavior could act as a mechanism for evolutionary change within a single species. Their models demonstrated that female choice could in principle allow some males to leave far more offspring than others, thereby driving the evolution of morphological and behavioral differences between the sexes. This theoretical return to sexual selection, in a Darwinian sense, was inspired by the earlier experimental results of the population geneticists.

These three synchronous stories intertwined during the 1950s and 1960s—the changing role of sympatric speciation models in evolutionary theory, the discovery and reception of the rare-male effect in France and the United States, and mathematical approaches to sexual selection in the later 1960s. Population geneticists insisted that theoretical mathematical models demonstrated the viability of female choice, and laboratory investigations demonstrated the efficacy of choice as a mechanism of evolutionary change. The question of whether female choice and sexual selection had measurable effects in wild populations of larger animals fell outside population geneticists' biological interests and professional jurisdiction.

Telling Apples from Oranges?
Theories of Speciation and Female Choice

The influence of the evolutionary synthesis research program of the 1940s codified species as "reproductively isolated units" and generated substantial interest among evolutionary biologists in how reproductive behavior affected species identity.[10] Biologists measured the degree of reproductive isolation by calculating the percentage of matings in which individuals chose a mate of the same species. By studying the effects of mate choice over time, they hoped to understand how species became reproductively isolated in nature. In particular, neo-Darwinian biologists sought to understand the dynamics of speciation and how isolation occurred, both geographically and physiologically. These biologists disagreed substantially, however, about whether to categorize "sexual or psychological isolation," in the form of "differences in scents, courtship behavior, sexual recognition signs, etc.," as a *result* of speciation or as a potential *cause* of speciation.

In the 1920s, scientists who argued for sympatric speciation through changes in reproductive behavior were predominantly sympathetic to the theory of mutation as a cause of evolutionary change. Gladwyn Kingsley Noble, for example, had described a hypothetical occurrence of speciation by psychological isolation as follows: "Female toads are attracted towards the male by his voice; if a change in the voice of a male should occur (as a mutational, a genetic difference), he probably would not attract a female, but if he should happen to seize a female, their offspring could only interbreed, for the voice of the offspring of the mutant would not be attractive to females of the original stock."[11] Because toads are typically nocturnal and mate under poor lighting conditions, female choice, in Noble's scenario, occurred through females' auditory preference for the calls of specific males. Once a female was close enough to a male to be "seized," her choice has been made. Noble's reliance on mutation as a normal part of speciation was not received particularly well by naturalists who wished to defend their domain against the incursion of geneticists. Henry Fairfield Osborn, honorary curator of vertebrate paleontology at the American Museum of Natural History (AMNH) and a professor of zoology at Columbia University, argued in response that speciation was "always adaptive," whereas "mutation is an abnormal and irregular mode of origin, which while not infrequently occurring in nature is not essentially an adaptive process; it is, rather, a disturbance of the regular course of speciation."[12] For this generation of curators at the museum, behavior as a mechanism for sympatric speciation became linked to evolution through mutation, a position that fell out of favor in the 1940s.

In 1942, Mayr, still a curator at the AMNH, outlined a process of speciation due primarily to geographic isolation.[13] He argued that geographic isolation was a necessary precursor for all speciation events. Only when two interbreeding populations were physically separated, he insisted, could evolutionary changes accumulate to such an extent that interbreeding would not occur when the populations again came into geographic contact. If sufficient time had not passed for the two populations to differentiate, then they would simply remerge. In rare cases, the process of reproductive isolation would have begun but would not be complete by the time the two populations came back into contact. In these cases, intermediate hybrids would produce fewer offspring than either parental strain (because they suffered from reduced fitness), thus creating a strong selective pressure favoring individuals that chose mates from their own parental population. Dobzhansky, who began working at Columbia University in 1940, called this period of renewed contact accompanied by selection against hybrid intermediates "reinforcement." Dobzhansky disagreed with Mayr about the frequency of reinforcement in the process of speciation. While Mayr contended such selection against hybrids was rare, Dobzhansky posited that natural selection for reproductive isolation (in other words, selection *against* intermediates) was a necessary component of speciation. After all, Dobzhansky noted, geographic isolation did not always lead to speciation, so some other factor must be involved in those cases that did lead to speciation.[14]

Both Mayr and Dobzhansky considered natural selection to be an important component of speciation, but in different ways. Mayr posited that speciation was an accidental by-product of natural selection acting independently on isolated populations. For example, the effects of natural selection would differ greatly in an isolated pocket of individuals in the mountains and a larger population of individuals in a valley. Over time, each population would become adapted to its unique ecosystem. Dobzhansky thought that for speciation to occur, natural selection had to act directly on interbreeding populations—hybrid individuals (one parent from each population) must leave fewer offspring than nonhybrid individuals (both parents from the same population). Although some genetic differences might accumulate in the isolated populations, the reason that populations stopped interbreeding, he believed, was selection for mechanisms to stop hybridization, including mate choice as a kind of psychological recognition.

Mayr sought to downplay the role of female choice as a mechanism of evolution, because of sexual selection's history as a potential mechanism for sympatric speciation. In 1947, he defined sympatric speciation as "a rapid, if not almost instantaneous, process of species formation,"[15] a kind of process that more closely resembled mutationist theories of evolutionary change than gradual Darwinian evolution. Mayr identified sympatric speciation as the principal obstacle blocking a synthesis between evolutionary biologists who argued that speciation was primarily a result of ecology (pro–sympatric speciation) and those who emphasized geography as the determining factor (anti–sympatric speciation).[16] Refusing to compromise, he suggested that biologists should dismiss sympatric speciation as a viable process of speciation. Sympatric speciation, he said, could only take place through assortative mating, which occurs when two types of individuals (say, yellow and orange in color) mate only with individuals of the same type (yellow with yellow, orange with orange). Mayr then contended that positive assortative mating was unlikely to have a disruptive evolutionary effect on a population—the strength of selection was simply too weak. Even if positive assortative mating did exist, it was extremely unlikely to be linked with habitat preference. Additionally, adaptation to local conditions was not likely to lead to disruptive selection unless isolation was already present. Most importantly, however, sympatric speciation seemed to require "preadaptation," a condition Mayr found simply farcical. He objected to the idea of "preadaptation" because it assumed a single mutation could produce a "drastic change in phenotype" (such as Noble's example of a change in the calling frequency of male frogs) that would then act as the basis on which selection could isolate the two populations (those individuals with the new trait and those without). Mayr insisted that such differences were the result of many small mutations that could only accumulate if the two populations were geographically (and therefore sexually) isolated from one another. In effect, Mayr equated naturalists' support of sympatric speciation with either a commitment to saltationist de Vriesian mutations as the source of evolutionary change or, worse yet, to Lamarckian inheritance of acquired characteristics.[17] He speculated that the only reason naturalists resorted to sympatric speciation was that they had misunderstood their observations of species distributions, leading them to falsely dismiss geographic speciation in some cases.[18]

Over the next three decades, Mayr continued to insist that all the significant steps toward reproductive isolation occurred while two populations were geographically isolated. At a 1960 symposium at the University of Texas on vertebrate speciation, for example, Mayr repeated his view that reproductive isolation was an accidental by-product of independent selection in isolated populations, estimating that 97 percent of the genetic differences between species were acquired during geographic isolation. Reinforcement might account for the final 3 percent, but that represented an insignificant contribution to the overall process of speciation.[19]

In contrast, Dobzhansky was far more sympathetic than Mayr to the idea that female mating behavior could act as an evolutionary mechanism in nature. Dobzhansky became strongly convinced of the importance of selection for reproductive isolation through his debates with geneticists Hermann Joseph Muller and Sewall

Wright over the importance of random mutations in the evolutionary process.[20] Muller and Wright argued that reproductive isolation would gradually develop between two populations without any natural selection (either for environmental specialization or for reproductive isolation), as a result of the accumulation of random genetic changes in each population. They called this accidental, nonselective model for evolutionary change (including speciation) "genetic drift." Dobzhansky, and people trained in his laboratory, believed that such "isolating genes" were "established by a process of selection which derived its original advantage from the achievement of the isolating effect itself."[21] Genetic differences controlling female mating behavior, for example, were themselves selected for during the process of reinforcement. As a result, Dobzhansky thought, sympatric speciation might occur if selection pressure for isolating genes were strong enough.[22]

In Dobzhansky's view, sexual selection and sexual isolation acted together to produce species-specific sexually dimorphic characters—female choice, in this context, was a continuum of interspecific and intraspecific mating preferences. Even so, he noted cautiously that he used terms such as *preference* and *discrimination* "merely to avoid circumlocutions," as he had no evidence that the power of discrimination rested solely with females, males, or even with both sexes. The major difficulty facing Darwin's theory of evolution, he noted, was the lack of observational data supporting the notion of competition for mates.[23] Dobzhansky further argued that animals' nest-building, courtship, and copulatory behaviors were too complex and species-specific to have accidental evolutionary origins. Nor did he want to invoke natural selection as the cause of these behaviors, because they appeared to have no effect on individual survival. He postulated that by modifying Darwin's theory of sexual selection, he could circumvent these difficulties, and advised his readers to "keep in mind the possibility that species specific courtship and mating habits and associated structures may serve as 'recognition marks' of the specific identity."[24] If mating behaviors and structures did play a role in sexual isolation, Dobzhansky suggested, they could possess a "survival value" and therefore could develop as part of the process of speciation. Despite his cautionary note, Dobzhansky's intent was clear: changes in reproductive behavior during mating could affect both the evolutionary future of a single species and the process of speciation. His interpretation (repeated by his students and research associates) differed significantly from Mayr's view of geographic isolation as the most important factor in speciation.

The major debates within evolutionary biology shifted between the 1940s, when Mayr formulated his geographic model of speciation, and the 1950s, with the rapid growth of experimental approaches to population and ecological genetics. As a result, some biologists mistakenly drew parallels between Mayr's geographic speciation theory and Muller and Wright's interest in genetic drift. This was far from what Mayr had intended. In a 1961 letter to Phil Sheppard, Mayr wrote: "Frankly, I am a little tired of having my 1954 paper cited as 'a theory of genetic drift' when the major objective of the paper was to prove that the genetic changes in peripherally isolated populations were due to selection and not due to drift."[25] Although Mayr believed that isolation was an important component of speciation, he did not think that speciation could take place with geographic isolation alone—there must be both geographic separation and natural selection in each independent population.

Mate choice as a mechanism for sympatric speciation in animals received its first experimental boost in the early 1960s.[26] John Thoday, the newly appointed Arthur Balfour Professor of Genetics at the University of Cambridge, and his graduate student, John Gibson, conducted a series of experiments designed to investigate the effects of disruptive selection in populations of fruit flies.[27] Starting with a single *Drosophila* population, they selected for two traits—high and low bristle count.[28] Bristles are hairlike structures found on the bodies of fruit flies (visible in the *Drosophila* illustration in chapter 3). Only flies with the most or fewest bristles were allowed to reproduce, but those individuals were allowed to interbreed freely. At the end of generations of selection, Thoday and Gibson obtained two reproductively isolated populations—one interbreeding population of individuals with low bristle count, and another with high bristle count. Sympatric speciation, they argued, had occurred.[29] Although they conducted their experiments in a laboratory, they claimed they had demonstrated the *possibility* of sympatric speciation in natural populations. "It is for the student of natural populations to determine whether and how often sympatric speciation occurs. However, sympatric speciation has frequently been supposed impossible on theoretical grounds, and these can no longer be regarded as sound."[30] This experiment drew immediate attention from geneticists, and textbooks now hail their experiment as an example of how strong selection in populations can produce remarkable effects.[31] Additionally, Thoday and Gibson readily admitted that they may have imposed a strength of selection unlikely in nature, yet they insisted that weaker disruptive selection along these lines would at least establish genetically distinct subgroups (stable polymorphisms) in a population, even if complete reproductive isolation were never achieved. The question remained, were such situations likely in nature?

Shortly thereafter, John Maynard Smith used mathematics to bolster the position that, if such stable polymorphisms existed in natural populations, then the continued emergence of mating behaviors in a population could lead to sympatric speciation.[32] Recall from chapter 4 that Maynard Smith became interested in sexual selection theory through his graduate work on *Drosophila* with J. B. S. Haldane and maintained his interest in animal behavior through his long-standing friendships with Aubrey Manning and Margaret Bastock. When Maynard Smith swapped his experimental work for theoretical investigations of evolutionary theory, he entered a community of mathematically inclined biologists trained by Haldane, E. B. Ford, and Ronald Fisher—a primarily U.K.-based group interested in mathematical population genetics and ecological genetics. Ecological geneticists, like population geneticists, were interested in how populations of organisms changed over time, but were also strongly committed to studying natural populations of organisms other than *Drosophila*.[33]

In his 1966 paper, Maynard Smith outlined the conditions sufficient to cause sympatric speciation. Initially, the population had to exist in a variable environment in which the environmental variations correlated with genetically distinct subpopulations. Then, selection acting on the mating process (either directly or indirectly) could reproductively isolate the subgroups without complete geographic isolation. The critical feature of this scenario, Maynard Smith argued, was this initial setup; without these starting conditions, sympatric speciation was not possible.[34] He suggested that whether or not his readers interpreted this as an argument for sympatric speciation *in nature* depended entirely on whether they considered stable genetic polymorphisms possible or entirely unlikely. If the conditions were unlikely, then biologists could relegate sympatric speciation to a trivial role in the origination of species. Maynard Smith was careful to conclude that his mathematics demonstrated only the *viability* of sympatric speciation; he made no claims about whether such a result should or should not be expected in nature.

Maynard Smith's sympatric speciation paper opened with an epigraph from Ernst Mayr's *Animal Species and Evolution*: "One would think that it should no longer be necessary to devote much time to this topic, but past experience permits one to predict that the issue will be raised again at regular intervals."[35] Indeed, Mayr proved correct. By the mid-1970s, even general discussions of modes of speciation included a section on sympatric species.[36] Population geneticists became more receptive to the idea of sympatric speciation because they increasingly accepted female choice and sexual selection as viable mechanisms for evolutionary change that could operate in ways that differed from natural selection. The close association of sympatric speciation and interspecific female choice grew out of Dobzhansky's descriptions of female choice as a continuum of mating behavior that extended from intraspecific preferences of a female for certain males within her species to interspecific choices of males of her own species over males of other species. This close association also meant that for population geneticists, theories of sympatric speciation and female choice were predicated on Dobzhansky's balance model of ecological and genetic diversity in natural populations.

The Rare-Male Effect:
Female Choice versus Natural Selection

In the 1960s, evolutionary biologists also began to shift the focus of their research from investigating behavior only within the context of speciation and species isolation to investigating behavior as a mechanism that could alter the genetic structure of a single population. In this context, female choice in fruit flies became imbricated in a debate between Dobzhansky and Muller about the genetic structure of populations. The "classical-balance" debate, as it is now known, hinged on what the genetic structure of a population looked like. According to Muller (whose position became known as the "classical" view), in a population subject to natural selection, the differential survival of individuals should act to slowly winnow the genetic diversity of the population. According to Dobzhansky (the "balance" view), the individuals who survive in a population subject to natural selection were typically the most genetically diverse individuals, so that even a population under strong selection would maintain its genetic diversity. Thus, the search was on among Dobzhansky's colleagues and friends for mechanisms that could act to maintain genetic diversity in a population—and female choice fit the bill.

To understand the stakes in the "classical-balance" debate between Muller and Dobzhansky, we must look a little farther.[37] Muller contended that each gene locus has an optimal, "normal," allelic state. In a population under selection, then, the most successful individuals would be those with two copies of the optimal allele for many of their genes (that is, they would be homozygous for many genes). As a result, natural selection would act over time to reduce variation within individuals and within populations. Dobzhansky, however, argued that there was no such thing as an optimal or normal allele. In a population under selection, the most successful individuals would be those that had two different alleles for many of their genes (that is, heterozygous at most of their gene loci). As a result, natural selection would act over time to maintain variation within individuals and within populations. Dobzhansky believed that heterozygous individuals had an advantage over their homozygous peers, because ecological environments were highly variable. In such highly variable environments, offspring of individuals with greater genetic diversity would have a greater chance of surviving in an environment that differed significantly from their parents' environment. High degrees of homozygosity amounted to overspecialization that would hurt the evolutionary future of a lineage in a variable environment.[38] Dobzhansky encapsulated these two positions as *classical* (to apply to Muller) and *balance* (to apply to himself), words that evolutionary biologists then used to refer to their respective positions. Dobzhansky's graduate students and collaborators sought to defend his point of view in this debate, and much of their research on sympatric speciation and female choice was undergirded by a belief in the natural genetic diversity of populations.

In this debate, fruit flies stood in for humans.[39] In *Mankind Evolving*, Dobzhansky argued that *Drosophila* could act as a useful model for human evolution and population structure because both groups of organisms exhibited high degrees of genetic variation and polymorphism. "Human variation presents a situation vastly more complex than that in the Drosophila fly . . . We may, however, use the latter as a paradigm to help elucidate the former."[40] For Dobzhansky, even if most animals were not polymorphic (and therefore *Drosophila* could not act as a practical model for these species), the most important animal species *was* polymorphic—humans. Moreover, Dobzhansky's unification of flies and humans based on the genetic similarity of their population structure allowed him to argue that Muller's eugenic proposals to ensure the future of humanity by preserving the sperm of society's geniuses were misguided for the same reasons that Muller's analysis of population structure in *Drosophila* failed: no single ideal genotype existed in either flies or humans.[41] Natural selection acted to preserve the genetic diversity of a population by spreading small differences over a large number of individuals, and this diversity was the key to humans' evolutionary success.[42]

When Dobzhansky's scientific collaborators defended female choice as an evolutionary mechanism maintaining the genetic diversity of a population, they were articulating an experimental defense of his vision of the importance of that variation in maintaining healthy populations (fly or human). To this end, population geneticists advanced several theories, including the "rare-male effect." If females preferred to mate with rare, exotic, or minority males (they used all of these terms), then the offspring from these couplings would also carry the rare alleles, and the genetic diversity of the population would be maintained. On the one hand, the rare-male effect was simply another case of frequency-dependent selection. The textbook example of this is mimicry. Consider two species of butterflies, one that is bitter and distasteful to birds and one that is delicious, but the two species look alike (both have big pink spots). If the frequency of the delicious butterflies is significantly lower than that of the bitter-tasting butterflies, then birds may learn to avoid eating all butterflies with big pink spots. If the frequency of the delicious butterflies increases, however, then the birds may not learn to avoid pink-spotted butterflies.

Natural selection (in the form of bird predation) varies according to the frequency of individuals (delicious or bitter-tasting) in the population. Similarly, the rare-male effect—female choice for male A—depended on the frequency of male A in the population. On the other hand, the rare-male effect was an example of a kind of selection that could counter the effects of natural selection, by maintaining rare male types in the population. Advocates of the rare-male effect referred to it as "sexual selection" for both of these reasons—it represented a deviation from random mating and it could act in opposition to natural selection.

Discussions surrounding the rare-male effect, by both advocates and critics, provide us with a gauge of contemporary evolutionary thought on how behavior could act as a mechanism for evolutionary change within a species. Critics of the rare-male effect at first argued that the effect was not due to female choice but was instead a result of differential male ability to stimulate females. As part of this argument, they posited that the fitness of organisms should not be construed restrictively as differential contribution to the next generation, but needed to take account of the general vigor of the individuals. Later critics (of the very early 1970s) shifted the target of their critique and contended that the rare-male effect was simply a construct of the artificial conditions of the laboratory and that *Drosophila* was not a representative organism. Until the rare-male effect could be reliably demonstrated in wild populations, they insisted, its potential importance as an evolutionary mechanism in nature was unknown. The changing criticisms leveled at the rare-male effect reflected major changes in evolutionary theory. Population geneticists became more comfortable discussing female choice and reproductive behavior as potential mechanisms for evolutionary change in *Drosophila*. They also increasingly voiced concerns over the previously accepted notion that fruit flies were sufficiently simple-minded to behave "naturally" in a laboratory setting, and perhaps they should not be considered "representative" organisms.

The first biologist to observe frequency-dependent mating was Claudine Petit. In 1946, after serving in the French Resistance, Petit returned to university and then began working as a graduate student in zoology at the CNRS, where she worked under the guidance of Georges Tessier. Tessier suggested that for her doctoral research, Petit investigate whether sexual selection occurred in the fruit fly *Drosophila melanogaster*. He believed that sexual selection and female choice had been greatly neglected by evolutionary biologists and thus constituted a perfect opportunity for research. Yet, after her first experiments, Tessier told Petit that her project was too hard and she could pick another topic. She insisted on continuing, at first as a side project, and eventually Tessier relented. Petit described sexual selection as "in the air" after Julian Huxley's 1938 papers on sexual selection in birds and Dobzhansky's and Mayr's papers in the mid-1940s on behavioral mechanisms of species isolation, and she was surprised that the subject was failing to attract further research attention.[43] Tessier and Petit dismissed Huxley's arguments against sexual selection as "short-sighted" and took them as a challenge to continue their research on female choice in fruit flies.[44] If Huxley's papers had any effect on population geneticists, Petit argued, it was to open up the question of how females chose mates and how that might influence the evolution of a population.

Petit discovered that when a mutant, *Bar*-eyed male *D. melanogaster* was surrounded by other *Bar*-eyed males, he was far less successful at acquiring a mate than if he was surrounded by wild-type, red-eyed males. Petit tested for and found a similar effect with *white*-eyed mutants. For her experiments, Petit used population cages designed by Tessier and Georges L'Heretier in the 1930s.[45] In each cage, she created a *synthetic* population (her word), consisting of five to seven hundred reproductively mature flies, in which the genetic composition of the population varied. She used *Bar*- and *white*-eyed strains because both traits were easy to spot in offspring, enabling her to calculate how successful each type of male had been through a visual inspection of the next generation. In her first set of experiments, she left the population of flies for five days, after which she isolated each female in a separate vial; when the eggs laid in the vial developed into adults (about eleven days later), Petit analyzed all the offspring to calculate the relative success of each kind of male present in the original population. In her second set of experiments, she removed and isolated the females from the synthetic population after only fifteen hours of contact. After mating, a female may reject subsequent attempted copulations for a period ranging from twenty-four hours to a few days. In the five-day experiments, there was a possibility that females had mated with more than one male. By removing the females after only fifteen hours, Petit ensured that each female had mated only once and that all the offspring counted were fertilized with sperm from a single male.

Petit called the rare-male effect an example of *selection sexuelle* in fruit flies. She reasoned that if males' mating success was consistent for a particular male genotype—that is, if mutant males always fathered the same number of offspring, regardless of the genetic composition of the population—then the effect should be attributed to natural selection. And, if mutant males were consistently less successful than wild-type males, then it would be safe to assume that the difference in males' mating success was related to male vigor more generally. However, because the mating success of mutant males varied, depending on the environment (in this case, the genetic composition of the population), the observed effect must be due to female choice. For Petit, one of the important consequences of her research was that

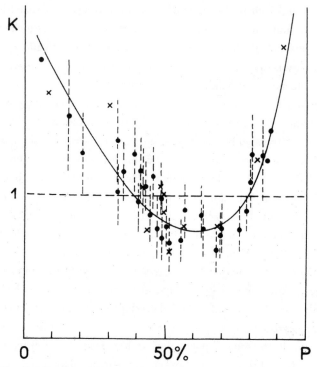

Claudine Petit's 1958 data on the mating success of *white*-eyed
flies in a population. The vertical (y) axis is K, the ratio of the per
capita mating success of *white*-eyed males to the per capita mating
success of wild-type males. The horizontal (x) axis is the percentage
of *white*-eyed males in the population, which ranges from 14 to 87
percent. According to the Hardy-Weinberg model of evolution,
if the ratio of mutant to wild-type individuals is the same in the
offspring generation as in the parental generation, evolution is not
occurring. In that case, the per capita mating success of mutant
and wild-type males should be equal, and K would hover near 1.
On this graph, that result would be evident as a horizontal line
(dashed line). The area at the upper left shows the rare-male effect:
when *white*-eyed males were relatively rare in the population, they
fathered a greater than expected proportion of offspring. From
Claudine Petit, "Le détermisme génétique et psycho-physiologique
de la competition sexuelle chez *Drosophila melanogaster*," *Bulletin
Biologique de la France et de la Belgique* 92 (1958): 269, fig. 3.

it demonstrated female choice and sexual selection in *Drosophila*. Also important to Petit, however, was that she could now explain the variation in scientists' measurements of the relative fitness of mutant strains of *Drosophila*—they had unwittingly been measuring the fitness of a single strain under different environmental conditions, so no wonder their numbers didn't agree. Fitness, she continued, fundamentally depended on the environment of the fruit fly and should not be ascribed solely to its genetic heritage.

Petit's work, published in French, remained largely unknown internationally and even within France. One of her first professional contacts outside France was Eliot Spiess. Spiess worked with Dobzhansky as a graduate student and earned his Ph.D. in 1949 from Harvard University, around the same time that Petit began her studies at the CNRS. In a surprising twist of fate, Spiess had been a member of the U.S. forces that entered Paris on August 25, 1944. He understood French, and he read Petit's papers with great interest. At the 1963 International Congress of Genetics in The Hague, he finally managed to track her down. He found her just as she was leaving the conference, and Petit remembers him shouting after her, "Madame Petit! Madame Petit!" They immediately became friends, and Spiess and his wife stayed in Paris an extra week, visiting Petit's laboratory and talking population genetics. In subsequent publications, Spiess cited Petit's papers as demonstrating the importance of regulating the frequencies of male flies in a population when calculating their relative fitness values.[46] If geneticists chose not to pay attention to the genetic composition of their populations, they might calculate radically different fitness values for the same fly type. Initially, then, Spiess considered the importance of Petit's work on the rare-male effect to be the identification of an important source of laboratory error.

Quite independently, another geneticist, Louis Levine, observed a similar effect in mice. At the time, Levine, who had received his Ph.D. under Dobzhansky at Columbia University in 1955, was a professor of biology at the City College of New York. He noticed in his breeding experiments that female preference for particular mates changed the gene frequencies of the population over time.[47] Much like Petit, he argued that sexual selection in populations might "explain the persistence of certain genes . . . in wild mice populations at frequencies that are in contradiction to the equilibrium frequencies postulated for them."[48] For population geneticists, a population "in equilibrium" meant a population that was not evolving—in theory, an infinitely large population in which all individuals mated randomly, with no mutation, no selection, and no migration (in or out of the population). In other words, if sexual selection were occurring, the population would not be mating randomly, the assumptions of an equilibrium population were not met, and the genetic composition of the population would change over time—evolution! This formulation created a role for sexual selection in population geneticists' models that was separate from the role of natural selection, as each type of evolution violated different assumptions of the Hardy-Weinberg equilibrium model. However, perhaps because he was working on mice rather than fruit flies, Levine's work did not garner significant attention within the drosophilist community until the 1970s.[49]

In 1965, several years after Petit's and Levine's initial work, Lee Ehrman also found that female fruit flies seemed to have a preference for mating with the rarest type of males in the population.[50] Ehrman, a former graduate student of Dobzhansky's and now working in his laboratory as a postdoctoral fellow, was investigating the effects of selection on the ability of *Drosophila pseudoobscura* to orient with respect to gravity. The fly variants in question did not mate as expected; female preferences for specific kinds of males varied with the frequency of those males in the overall population. Such sexual selection for rare males, Ehrman argued, meant that male evolutionary fitness varied with the environment.

Believing this was the first time anyone had noticed such a phenomenon, Ehrman excitedly ran into Dobzhansky's office, only to be informed by Ernst Bösiger (a French geneticist who happened to be collaborating with Dobzhansky in New York), subsequently confirmed by Eliot Spiess, that Claudine Petit had already discovered *and* published on the topic. The only difference between their findings was the species of fruit fly—Petit worked on *D. melanogaster*, and Ehrman had been studying *D. pseudoobscura* and *D. persimilis*. Ehrman published her findings in the U.S. journal *Evolution*, cited Petit's work, and thereby made Petit's research known to an English-speaking audience.[51]

Bösiger had already arranged for Dobzhansky to travel to France for research collaboration. Once he arrived in Paris, Dobzhansky sought out Petit, and she fondly remembers their chats about genetics over dinner in her apartment.[52] After his trip, Dobzhansky arranged for Petit and her daughter (then eighteen and working as her mother's laboratory technician) to come to the Rockefeller Institute in New York for a three-month collaboration with Lee Ehrman. During this three-month period, Petit became acquainted with several other international visitors, and she later traveled to Japan, Brazil, and Italy to collaborate with drosophilists she met in New York. Petit and Ehrman collaborated on several publications over the next few years, while also continuing to publish independently. They identified the importance of the rare-male effect not as a source of experimental error, as had Spiess, but as demonstrating Darwinian female choice and sexual selection in fruit flies, which Levine had seen as the importance of his results in mice.

The reception of the rare-male effect by the population genetics community highlighted the importance of polymorphisms and the genetic structure of populations in evolutionary theory of the 1960s. Several questions arose. First, what was *sexual selection*—was it simply any deviation from random mating in a population, or was it a particular mode of female evaluation and choice of mates? Second, what was meant by *fitness*—was it merely reproductive success, or did it imply something about the physical state of the organism? Third, what was the meaning of *natural*—were Petit's and Ehrman's experiments artificial because they took place in a laboratory and because they used *Drosophila* as their sole experimental organisms, or did their experiments actually demonstrate something about how evolution acted in wild populations in nature?

For Ehrman and Petit, the importance of the rare-male effect ultimately rested in the potential of this kind of sexual selection to maintain the genetic heterozygosity of natural populations. They used their results to defend Dobzhansky's balance model of genetically variable natural populations. Both argued that female choice for rare males would act to maintain genetic diversity by preventing complete elimination of rare genotypes from a population.

Other biologists questioned Petit's and Ehrman's correlation of female choice and genetically diverse populations and argued that populations with high degrees of heterozygosity could be maintained through other means. For example, Bösiger suggested that the heterozygote effect through male-male competition could also explain high degrees of genetic diversity. He argued that the genetic diversity of a population did not have to be maintained through many individuals that differed at a few gene loci. Instead, genetic diversity could be maintained if a few individuals that contained a great amount of genetic variation were more successful than others at obtaining mates.[53] As evidence, he performed several experiments looking at the overall success of wild-type and mutant males of *D. melanogaster* in a single population. His wild-type males contained more heterozygous alleles than their mutant (homozygous) brethren. He placed a single male in a cage with a variety of females and, after waiting forty-eight hours, dissected the females to find what percentage had been inseminated. Bösiger measured reproductive success of males in terms of copulation rates rather than by analyzing genetic contribution to the next generation, as had Petit and Ehrman.

Bösiger cited Petit's and Ehrman's papers not as demonstrating a rare-male effect, but as providing evidence that, in most cases, wild-type males were more successful than mutant males (which, in total proportions, they were). For him, Petit and Ehrman's conclusion was correct—sexual selection in a population would lead to increased heterozygosity—but their proposed mechanism (female choice) was wrong. This strong case of sexual selection, he argued, was "mainly, if not exclusively, a consequence of differences in the sexual vigour of males and the receptivity of females." Bösiger saw his research as supplying additional evidence in support of Alfred Sturtevant's conclusions in his 1915 paper on sexual selection in *Drosophila*: female flies didn't choose; but the sexual vigor of males was determined by their genetic endowment.[54]

Bösiger's interpretation revealed another fundamental difference in the way he understood mating behavior. Ehrman and Petit advocated a choice-based model, in which female flies preferred to mate with some kinds of males more than others. Bösiger's model of mate choice relied on male vigor as an explanation for Ehrman's and Petit's data; apparent female "choice" was due to variation in the amount of stimulation a male produced during his attempts to induce females to mate. More vigorous males were better at providing that stimulation and were therefore able to mate with a greater number of females. A female, in Bösiger's model, would mate with the first male to stimulate her sufficiently. His interpretation linked closely the concept of male-male competition and the specificity of courtship signals at a species level: only healthy males of the right species would be able to adequately stimulate a female to induce a sexual response (an argument reminiscent of Julian Huxley's convictions about the function of courtship).

Yet, despite these very real differences in interpretation, all three geneticists—Petit, Ehrman, and Bösiger—agreed that sexual selection, whether construed as female choice or male-male competition, could increase the genetic diversity of a population. They all advanced the idea that sexual selection was a mechanism of evolutionary change that could counter the winnowing effects of natural selection on genetic diversity.

For other geneticists, as well, the rare-male effect reinforced the idea that sexual selection and sexual isolation were part of the same continuum, a model Dobzhansky had proposed in 1944—though few biologists made note of his article when it was first published. By 1971, A. Faugeres, one of Petit's graduate students at the Laboratoire de Génétique des Populations, Université de Paris, argued that sexual selection and sexual isolation were two aspects of the same evolutionary phenomenon. "Sexual isolation," she continued, "is only a particular case of sexual selection, affecting both sexes and thus causing the convergence of sexual choices, and it seems quite logical to include under the general term 'sexual selection' all mechanisms that cause deviations from panmixia [random mating]."[55] In this continuum, all forms of nonrandom mating counted as sexual selection. At one end, positive assortative mating (like forms having an increased tendency to mate with each other) would lead to sexual isolation; at the other end, frequency-dependent mating (such as the rare-male effect) would lead to a genetically balanced population in which rare types were maintained.

The reception of this spin on Petit's and Ehrman's research revealed a second tension within the evolutionary community. Was a male's mating success dependent on his environment? In other words, was fitness relative, in that it was only possible to measure the fitness of an individual with respect to a specific environment and population composition? Or was fitness absolute, in the sense that each male possessed a finite amount of vigor?

In 1963, two years before Ehrman rediscovered Petit's rare-male effect, Julian Huxley had published a book review of Dobzhansky's *Mankind Evolving*.[56] Huxley argued that the biological community needed to consistently distinguish between "reproductive" and "survival" selection. Fitness, he continued, could be conceptualized as either physical vigor or the ability to have many offspring, yet biologists were not consistent in using the term one way or the other and this had led to a great deal of confusion in the evolutionary literature. The crux of his anxiety was the tendency of population geneticists (especially Dobzhansky) to assume that fitness was *solely* reproductive success and not vigor or health more generally. Huxley then wrote several letters to his friends, asking them for their definitions of fitness and advancing his own.[57] Most of his correspondents were receptive to Huxley's line of thought. For example, Ernst Mayr agreed completely, writing in response to Huxley's missive: "You may have noticed that in my discussion of natural selection, I very emphatically separated the advantage given by *genuine* fitness from the advantage given by *mere* reproductive success. This is precisely the point you are making." Two weeks later, Bernard Rensch responded as well. "I must confess that I was also astonished to see that both Dobzhansky and Simpson used this term [fitness] only in the sense of reproductive advantage. Well, Dobzhansky is a geneticist and he therefore mainly sees this genetically measurable character. Simpson, however, could have seen that such a one-sided definition does not cover the meaning of the word and especially not in the sense of Darwin."[58]

The connection between definitions of fitness and levels of selection compounded the issue even further. Huxley and Mayr argued that the individual was the unit on which selection acted, and therefore only individuals could have a true "fitness value." For geneticists, however, fitness could be the property of an individual in a population of individuals, or of an allele in a population of alleles. And therein lay the problem: which really counted, genes or individuals? Bösiger's interpretation of Petit's and Ehrman's data as revealing the importance of male vigor, not female choice, built on a tradition of fitness conceptualized as a real physical phenomenon residing in the bodies of the flies. For Petit and Ehrman, fitness was simply a statistical phenomenon, and one that depended on the environment.

Most population geneticists, in fact, agreed with Petit and Ehrman in their interpretation of fitness as relative. After reading Huxley's review of his book, Dobzhansky wrote Huxley a letter in which he responded that "the concepts of natural selection and of fitness have evolved and changed since Darwin and Spencer; so have other fruitful concepts in science. Our concept of species is not identical with that of Linnaeus, and of gene with that of Mendel or Bateson. Owners of Greek dictionaries may, of course, manufacture new and fresh terms. I do not find this either expedient or attractive."[59] Dobzhansky stuck to his belief that true fitness was reproductive success and implied nothing about the physical vigor of the individual. Another population geneticist, Ken-Ichi Kojima, of the University of Texas at Austin, published a paper in 1971 entitled "Is There a Constant Fitness Value for a Given Genotype? No!"[60] Kojima cited Petit's and Ehrman's work as providing the first *experimental* evidence of frequency-dependent mating selection. His conclusion was as emphatic as his title: "Let it stand that non-constant fitness is a rule in population studies." Kojima argued that Petit's and Ehrman's work clearly demonstrated how populations could evolve in response to differential reproductive success. Additionally, he suggested, their work showed that the reproductive success of individuals varied in different environmental contexts. Thus, female choice (sexual selection) would act to maintain the genetic variation of individuals and of the population more generally. By 1975, evolutionary theorists posited that the rare-male effect constituted the only known mechanism by which sexual selection could act to maintain the genetic diversity of a population.[61]

Despite the demonstrated efficacy of sexual selection in laboratory populations of flies, some biologists expressed concerns that the rare-male effect would never be demonstrated in organisms other than *Drosophila*, or outside a laboratory. In 1970, for example, Pomeroy Sinnock published the results of a set of experiments designed to address at least the first of these concerns. For his graduate work in genetics at Berkeley, working under I. Michael Lerner (another friend of Dobzhansky's), Sinnock demonstrated the existence of a minority effect in the flour beetle, *Tribolium castaneum*. Yet he despaired over estimating the importance of his artificial experiments for understanding evolution "in nature."[62] Similar attempts to discover the rare-male effect in the common house fly failed, but a minority effect was observed in a small parasitic wasp.[63] Criticisms of the rare-male effect and frequency-dependent selection as being potentially limited to *Drosophila* reflected the increasing skepticism of organismal biologists toward the laboratory environment as the ideal place to study evolution in action.

In summary, then, advocates of the rare-male effect associated female choice with sexual selection, both because it represented a deviation from random mating and because they saw it as acting in opposition to the potentially winnowing effects of natural selection on the genetic diversity of a population. Rather than searching for ways in which female choice could act as a mechanism for speciation (as had John Thoday and John Maynard Smith), Claudine Petit and Lee Ehrman strengthened population geneticists' conviction that female choice could act as a powerful mechanism of selection within a single population, at least on paper and in the laboratory.

Sexual Selection and Theoretical Population Genetics

Once experimental population geneticists accepted female choice as a viable mechanism of selection within a species, theoretical population geneticists began to extend their models to address the question of whether female choice could also lead to sexual dimorphism. As a line of thought, why males and females looked different had been far from population geneticists' research in recent decades. When theoretical biologists became interested in female choice as a mechanism of evolution in the late 1960s and early 1970s, they built on the experimental and theoretical conclusions of drosophilists: females were both theoretically and practically capable of choosing their mates. The research of experimental population geneticists on evolution and female mate choice would eventually filter into organismal biology through these theoretical models of sexual selection developed in the 1960s.

Theoretical biologist Peter O'Donald, for example, used computers to model the process of sexual selection for his doctoral research in the Department of Zoology at the University College of North Wales, Bangor.[64] O'Donald believed, as had Darwin and Fisher before him, that individuals who bred early in the reproductive season would be favored by both natural and sexual selection.[65] Presumably, male birds that bred early in the season must be more attractive than males breeding later, because females had chosen them first. If these birds also produced more offspring than later breeders, then the effects of female choice and natural selection would be mutually reinforcing. To test the model, O'Donald gathered data on the breeding of arctic skuas (piratical seabirds that nest in the tundra) and found that, yes, earlier breeders did produce greater numbers of offspring.[66] So, in the case of breeding time, attractiveness correlated with higher fitness (reproductive capacity) and therefore was a good indication of mate quality.

The payoff from this arctic skua research was O'Donald's demonstration that sexual selection could operate in monogamous populations of birds in addition to polygamous populations. For monogamous species, variation in breeding times correlated with fitness could lead to stable polymorphisms in a population, in the same way that the rare-male effect could act in polygamous species.[67] Few people seemed interested in his sexual selection work, however, and he began to feel despondent about his academic future.[68] His friendship with British ecological geneticist Philip Sheppard kept him going. With Sheppard's urging, he began to work on the evolution of mimicry, and not until the 1970s did he return full-time to the question of sexual selection.[69]

O'Donald now concentrated on Fisher's theory of runaway sexual selection. Runaway selection differed from the Darwinian model in that in Fisher's theory, females preferred arbitrary traits in males. Males possessing the exaggerated trait would be more attractive to females and leave more offspring *merely* because of their beauty. "Fisher's 'runaway process of sexual selection,'" O'Donald insisted, ". . . stops only when the preferred males have acquired so great a disadvantage by natural selection as to outweigh the sexual selection."[70] He explored Fisher's theory as an example of an evolutionary mechanism that could run counter to natural selection.

Although he drew heavily from the population genetics research of Ehrman, Petit, and Spiess, as well as Maynard Smith's earlier work, O'Donald construed female choice in ethological terms. He mathematically modeled female choice as "the likelihood that the males' display will release [females'] mating responses."[71] The rare-male effect, or female *choice*, he defined as females' possessing lower behavioral-release thresholds for certain kinds of males.[72] For O'Donald, insisting that females were choosing amounted to no more than assuming "that females have certain thresholds in their receptivity to male courtship that must be exceeded before they mate," even in birds.[73] By framing female choice in this way, O'Donald could parameterize choice in his models and investigate the evolutionary results of stronger and weaker preferences in female mating behavior. Yet in doing so, O'Donald redefined drosophilists' vision of active choice as passive threshold.

As the currency of theoretical models of sexual selection in biology increased in the 1970s, *Drosophila* simultaneously began to lose its representative status among biologists interested in animal behavior. In an interview with John Maynard Smith in 2002, I asked whether modern ornithologists were aware of this behavioral research in *Drosophila*. He responded by saying, "No, I don't think they are. *Drosophila*'s not an organism, to them." In a similar conversation with Lee Ehrman, she exclaimed: "Imagine if we could go into the field and see our *Drosophila* mating. I mean if we could actually *see* that! There have been a few people who have done that—have *tried* to do that. Sometimes [fruit flies] will come into your buckets, into your baits—see it's already unnatural—and mate there . . . It's just a matter of vision, I think. The main data gathering sensory organ[s] of humans are our eyes . . . and you can't see [fruit flies] in the field, you really can't."[74] Maynard Smith's articulation of a divide between laboratory and field research on animal behavior and Ehrman's dismay at the lack of field research on *Drosophila* behavior both speak to an increasing appreciation among biologists for the variability of fruit fly behavior in different environmental contexts.

Over the 1950s and 1960s, zoologists had become increasingly aware that *Drosophila* species were organisms in which the expression of particular behavioral patterns depended on the environmental conditions of the laboratory. The rare-male effect demonstrated that preferences exhibited by females for a particular kind of male changed according to how frequently the female encountered such a male in the population. Additionally, researchers began to perform "conditioning" experiments with *Drosophila* and discovered that fruit flies could be trained to change their reactions to different scents.[75] Later research further demonstrated that the conditions in which flies were raised could affect their behavior as adults.[76] Discussions in these papers emphasized the importance of regulating experimental conditions to prevent aberrant behavioral responses. The problem was that nobody knew what normal environmental conditions were for *Drosophila*, nor could they easily find out.

In 1970, geneticists interested in behavioral studies founded an independent journal in which to publish their research, *Behavior Genetics*. This journal provided a venue for genetic investigations of behavior and was characterized by research on fruit flies, mice, and humans. The journal also marked an increasing disciplinary divide between geneticists and organismal biologists over how to best study the evolution of behavior.

In the post-synthesis theoretical and population genetics community of the 1960s, the relationship between female choice and speciation had driven much of the research on sexual selection as a mechanism of evolution and helped solidify the position of sexual selection within the theoretical discourse of the life sciences. Yet organismal biologists of the 1970s rarely cited the work of Ehrman, Petit, and Spiess, even though their experiments on female mate choice in *Drosophila* were crucial to the initial development of theoretical models of female choice and sexual selection. Instead, organismal interest in female choice and sexual selection derived from models of evolutionary change developed by evolutionary theorists such as John Maynard Smith and Peter O'Donald. The connection of these mathematical theories to experimental population genetics had been severed.

By the end of the 1960s, theoretical and experimental population geneticists were no longer concerned with the question of whether or not a female could choose. Experimental evidence demonstrated to their satisfaction that females distinguished between males of their own species and of different species and preferred to mate with some males of their own species more than others. A female's choice in mates, therefore, could range from interspecific to intraspecific preferences.

The issues of sympatric speciation, the rare-male effect, and sexual selection were all part of the same nexus of problems for theoretical and experimental population geneticists of the 1960s. Within the context of reproductive behavior, these biologists were primarily interested in understanding the evolutionary effects of female choice on the genetic architecture of nonhuman animal populations and how that structure would change over time as a result of mating preferences. Population geneticists interested in the effects of reproductive behavior on the evolutionary process gave renewed vigor to a vision of female choice as a mechanism of evolutionary change in terms of both speciation and intraspecific evolution. However, research on female choice in the 1960s also revealed evidence of deep philosophical divisions among evolutionary biologists over such issues as the role of behavior in speciation, the definition of fitness, the potential importance of group-level processes in evolution, and the relevance of laboratory research on insects to understanding the process of evolution in nature.

At a very basic level, Ernst Mayr doubted the importance of female choice as an evolutionary mechanism. He argued that if female choice were important only within the context of sympatric speciation and as a mechanism to increase or maintain the genetic diversity of a single population, then it could not be a very significant evolutionary mechanism in explaining the diversity of species (even if it did operate in some populations). Yet most of the *Drosophila* geneticists with an interest in mating behavior were associated with a different architect of the synthesis, Theodosius Dobzhansky, either as graduate students or as visiting researchers and collaborators. Inspiring long-lasting friendships with many of the students and researchers who passed through his laboratory, Dobzhansky collected an international community of geneticists interested in behavior as a mechanism for both sympatric speciation and evolution within a single species or population.[77]

By recognizing female choice and sexual selection as viable evolutionary mechanisms, these experimental and theoretical population geneticists created a conceptual space for evolutionary processes that could act independently from natural selection. Sexual selection as a mechanism for speciation could result in sympatric speciation. Female choice for rare males, as a mechanism for changing gene frequencies in a single population, acted to preserve the genetic diversity of the population, even as natural selection acted to cull less fit individuals from the breeding pool. Female choice as a mechanism for morphological change in a species could cause exaggerated male traits even though the males who exhibited them were more susceptible to predation. Female choice, then, could act both in concert with natural selection (through sympatric speciation or the rare-male effect) and in opposition to natural selection (as Darwinian sexual selection).

Most models of female choice in population genetics also presumed that mate-choice decisions were for the evolutionary benefit of both the individual making the choice and the species as a whole. This was especially true for the rare-male effect. In the rare-male model, population geneticists posited that female choice acted to maintain the genetic diversity of the population. The evolutionary advantage of such genetic diversity rested in the possibility that a variable population was less likely to become extinct in variable environmental conditions. In the context of the rare-male effect, the ability of a species-level advantage to drive the evolution of individual behaviors maintained its currency into the mid-1970s, far after the ghost of group selection was ostensibly vanquished by George Williams's 1966 book *Adaptation and Natural Selection*.[78]

Another issue that research on female choice rendered especially visible was the argument over definitions of organismal fitness. In the mid-1960s, Julian Huxley instigated a movement to distinguish between reproductive fitness and survival fitness. Although some members of the evolutionary biology community had been claiming this for years (Simpson and Mayr, as well as Huxley, for example), with the increasing authority of molecular biology following the discovery of the structure of DNA in 1953 and the decoding of DNA in the 1960s, a greater number of biologists began to insist that population geneticists had inappropriately fused reproductive and survival success in their mathematical descriptions of how evolution progressed in a population.

In most cases, geneticists carefully separated the "scientific" results of their experiments from the "popular" implications for human evolution, a common strategy by mid-century. Over the course of the twentieth century, biologists had increasingly differentiated between publications intended for a popular audience, in which they deemed it acceptable to talk about the implications of research on animals for understanding people, and those intended for a professional audience of biologists, in which research on animals was intended to be an end in itself. In professional publications, discussing humans might have been interpreted by other biologists as unprofessional, anthropomorphic, or simply overstepping the logical bounds of scientific discourse.[79] Although population geneticists typically saw their experiments on flies and other insects as relevant to understanding the genetics of humans (this was especially clear in the debates between Dobzhansky and Muller, for example), they rarely said so in their scientific articles, preferring to reserve such comments for books and philosophical musings. These more popular venues, however, helped secure a firm place for sexual selection in evolutionary theory starting in the 1970s.

CHAPTER 6

Selective History:

Writing Female Choice into Organismal Biology

In a way the impact of biology has had a curious set-back as a result of the magnificent victories of molecular biology. To the outsider they suggest that physics and chemistry is the A[lpha] and O[mega] of all science. We will have to make a double effort to restore the influence of organismic biology and to make better known the evolutionary trends that culminated in that unique psycho-social organism Man.

—Ernst Mayr to Julian Huxley, October 3, 1967

By the 1960s, biologists could no longer say with certainty that only humans used tools, or could learn to communicate through language, or behaved according to rational rules.[1] Jane Goodall's discovery that chimpanzees used and manufactured tools surprised the scientific community. On hearing the news, anthropologist Louis Leakey, one of Goodall's mentors, remarked: "Now we must redefine tool, redefine Man, or accept chimpanzees as human."[2] Throughout the late 1960s and early 1970s, Goodall's work received wide circulation through her articles in *National Geographic* magazine, the documentary *Miss Goodall and the Wild Chimpanzees*, and her best-selling book *In the Shadow of Man*. Goodall was young, white, beautiful, and feminine. According to feminist historian of science Donna Haraway, in the eyes of eager watchers, Goodall symbolically mediated the gap between white Western civilization and the primitive beasts of the Dark Continent. Goodall's movie and associated articles in *National Geographic* magazine were a huge success.[3] Science bridged the connection between human and animal as Goodall traipsed through the trees and over the hills of African woodlands, a thin sprite at home in her tropical Eden. More exciting news came in 1969, when primatologists R. Allen Gardner and Beatrice T. Gardner described in *Science* how they had taught a young female chimpanzee named Washoe to communicate using American Sign Language.[4] The rising popularity of animal behavior research in the popular media, and the success of animal stars in the movies and television shows of the 1960s and 1970s, also facilitated an easy slippage between human and animal social and sexual behavior.[5]

Then, in the 1970s, theoretical biologist John Maynard Smith and geneticist George R. Price published a series of papers, including "The Logic of Animal Conflict."[6] They mathematically described the evolutionary success of retaliation, escalation, and conflict avoidance in long-term warfare scenarios in which animals could learn from their past encounters with other participants. Maynard Smith and Price dubbed the unbeatable animal strategy for each set of conditions the "evolutionarily stable strategy" for that "game." Evolutionary biologists quickly adopted these analytical tools, calling them "evolutionary game theory." Animals' choices in these games were determined by their genetically preprogrammed strategies. Animals with better strategies survived and produced more offspring. The rise of game theory in biological circles gave biologists new license to discuss apparently rational behavior in animals, especially choice-based behaviors, without the anthropomorphic connotations of aesthetic or cognitive evaluation and decision-making. By 1976, evolutionary biologist Richard Dawkins hailed Maynard Smith and Price's paper as one of the most important contributions to evolutionary theory since Darwin.[7]

All told, these developments removed tool-manufacture, sociality, the mental capacity for language, and even apparent rationality as physical or behavioral signals demarcating a hard and fast boundary between human and animal. These developments were especially important for biologists whose research centered on levels of biological organization higher than subcellular processes—in effect, those biologists who worked with whole organisms and populations. Theodosius Dobzhansky wrote to a friend that "organismic biology is fundamental for the understanding of man, while molecular biology is more important for the understanding of his ailments, and for finding a cure against colds."[8] The hostility in his letter reflected a general concern among organismal biologists about molecular biology's rising authority as the final frontier of biological research. Ernst Mayr, for example, agreed that insights into the human predicament made organismal biology as important as molecular biology: "in addition to that wonderful field of molecular biology, we have an equally wonderful field of organismic biology, a field which is becoming increasingly important for the understanding of man and the planning of his future."[9] Female choice once again made news in the popular press in the context of organismal biologists' research on sexual selection as applied to animals and, ultimately, humans.

When reading through back issues of *Evolution, Animal Behavior, Trends in Ecology and Evolution,* or even the *Journal of Theoretical Biology,* one cannot escape the feeling of excitement accompanying the explosion of interest in female choice beginning in the 1970s. Rocketed to scholarly and public attention by Edward Osborne Wilson's *Sociobiology: The New Synthesis* (1975) and Richard Dawkins's *The Selfish Gene* (1976), sexual selection seemed to provide a reconceived theoretical basis for the Darwinian analysis of human and animal sex differences.[10] Subsequent research by organismal biologists took for granted that female animals were capable of "choosing" their mates (in a game-theory sense, at least), but continued to question the basis of that choice: was a female's choice arbitrary (as supposed by Ronald Fisher's theory of runaway sexual selection), were females judging the genetic quality of males through the expression of a male's exaggerated trait, or did the trait itself correlate with some direct benefit to the female (such as a high-quality territory)?[11]

After the excitement of this moment, perhaps it is not surprising that according to most histories of sexual selection written in the 1980s, biologists' interest in female choice seemed to have undergone a period of neglect for much of the twentieth century. The "eclipse narrative" of the history of sexual selection quickly gained traction within two disparate academic conversations: the importance of organism-centered research in the biological sciences and the feminist critique of science.

For organismal biologists, the eclipse narrative demarcated sexual selection as an experimental field science (well outside the jurisdiction of molecular biology). In response to molecular biologists' pointed challenge—What theories had zoologists produced lately that were new and interesting?—sexual selection provided a robust answer. Sexual selection was a promising line of research in the 1970s; it was a relatively new research focus for organismal biologists; it seemed to have consequences for scientists' understanding of human behavior; and, perhaps most important, sexual selection was best studied in wild species. Molecules could not provide the *sole* key to the future of biological research. As an increasingly popular and controversial theory, sexual selection became the lens through which many organismal biologists analyzed sex differences in appearance and behavior in animals and humans.

The eclipse narrative also became important in second-wave feminism's critiques of evolutionary theory.[12] Appropriating the eclipse narrative of sociobiological texts, feminists suggested that sexual selection had been ignored as a theory for the better part of a century because *female* choice was (however unconsciously) a difficult pill for predominantly male biologists to swallow.[13] Female choice, then, according to this interpretation, could gain acceptance within the biological community only after the feminist movement of the 1960s had normalized the issue of women's active role in society. Feminists also critiqued contemporary sexual selection theory as having uncritically imported sexual stereotypes directly from the time of Darwin to the present—coy females and eager males.[14] The arguments in sociobiological analyses of sexual behavior, feminist narratives generally stressed, reflected the cultural assumptions of biologists, not universal scientific truths.

The arguments of both biologists and feminists in the 1980s were strengthened by the apparent eclipse in biologists' interest in female choice for the better part of the twentieth century. Yet, as we have seen in the previous chapters, biologists' interest in female choice never truly disappeared. Population geneticists and quantitative theoretical biologists had been studying female choice for decades. In this chapter, I explore first how organismal biologists came to embrace sexual selection—for good, this time—as an important mechanism of evolutionary change in nature, then the iterative process by which histories of sexual selection and female choice proffered by different communities of biologists stabilized into a single historical account: the eclipse narrative. Although the eclipse narrative of sexual selection that emerged during this period functioned brilliantly as a device demarcating the disciplinary identity of sexual selection as an organismal rather than genetic program, it failed to yield a nuanced history of female choice in evolutionary biology.

(Re)theorizing Female Choice

When I interviewed Robert Trivers in 2004, I met him in downtown New Brunswick, New Jersey, at a sushi restaurant close to Rutgers University, where he is on the faculty of the Anthropology Department. Prone to periods of intense activity and depression throughout his life, that day he was *on*. Over the course of the next several hours, he never stopped moving, never stopped talking. He spoke with his usual candor about his past and clearly relished his reputation as a maverick within the scientific community. Trivers said that he first learned about evolution in the 1960s by working on a series of curriculum guides for elementary school children, called "Man: A Course of Study" (MACOS). Part of the American educational reforms that followed the Soviet launch of Sputnik, MACOS was designed to improve the quality of social science education in the classroom. Trivers translated the primatological research of Irven DeVore on baboons in Amboseli National Park, Kenya, into language the school children could understand. From the beginning of his biological education, Trivers had thought of animal behavior and evolutionary theory in terms of understanding humans.[15]

By 1972, Trivers was finishing his graduate work at Harvard University, and after earning his Ph.D., he immediately started a position there as a junior faculty member in the Department of Anthropology. Both DeVore and Ernst Mayr have discussed their roles as mentors and advisors to Trivers during his Harvard years.[16] Although DeVore was Trivers's official academic advisor, Mayr would later claim that "Bob

Trivers, who was as much my student as anybody's, developed his important chapter [in Bernard Campbell's *Sexual Selection and the Descent of Man, 1871–1971*] in frequent conversations with me"—testament, perhaps, both to Trivers's intellectual promise and to his need for personal guidance.[17]

Between 1971 and 1974, Trivers wrote four papers that have become classics in the application of evolutionary theory to behavior—from reciprocal altruism to parent-offspring conflict and sexual selection.[18] He fondly remembers these years at Harvard, when he flew "under everyone's radar" and remained relatively unknown.[19] After the publication of Wilson's *Sociobiology*, Trivers has said, his ideas became very popular very quickly, and "the proverbial you-know-what hit the fan."[20] His early papers have remained influential for many evolutionary biologists, and in 2007 Trivers was awarded the prestigious Crafoord Prize in Biosciences for his "pioneering ideas on the evolution of the social behaviour of animals." The Crafoord Prize is the Royal Swedish Academy of Sciences' recognition of fields in which it is impossible to earn a Nobel Prize, chiefly mathematics, geosciences, and nonmedical biosciences.

Trivers's 1972 paper "Parental Investment and Sexual Selection" appeared as a chapter in anthropologist Bernard Campbell's *Sexual Selection and the Descent of Man, 1871–1971*, a volume celebrating the centennial of Darwin's *Descent of Man*.[21] Based on a recommendation from Mayr, Campbell asked Trivers to write a review of sexual selection in lower organisms.[22] Although Trivers did include a review of previous work in his chapter, he devoted the bulk of his efforts to providing an explanation for female choice based on differential parental investment. Having discovered A. J. Bateman's research on mate choice in fruit flies while reading genetics papers with Mayr as part of his requirements for graduate school, Trivers argued that sexual selection was likely to occur in any species where the male and female contributed disproportionately to the care of their offspring.[23] He reasoned that if females only mated once per mating season, they would invest substantial energetic resources in (depending on the kind of organism) producing eggs, giving birth, and/or raising offspring. They would also be extraordinarily picky in their choice of mating partner. Further, Trivers continued, females in a population varied little in the number of offspring they could produce, so the main difference between a successful and unsuccessful female must rest in the quality of her offspring, which was determined in part by the quality of her mate. All of this assumed, of course, that the male contributed very little to the energetic cost or upkeep of his offspring, allowing him to mate with a large number of females. His mating success then depended on the quantity, not the quality, of his offspring. As males could mate multiple times per season, reasoned Trivers, they probably invested very little in any

Robert Trivers's Theoretical Criteria for Female Choice of Males

I. All species, but especially those showing little or no parental investment
  A. Ability to fertilize eggs
    (1) correct species
    (2) correct sex
    (3) mature
    (4) sexually competent
  B. Quality of genes
    (1) ability of genes to survive
    (2) reproductive ability of genes
    (3) complementarity of genes
II. Only those species showing male parental investment
  C. Quality of parental care
    (1) willingness of male to invest
    (2) ability of male to invest
    (3) complementarity of parental attributes

Source: Robert L. Trivers, "Parental Investment and Sexual Selection," in *Sexual Selection and the Descent of Man, 1871–1971*, ed. Bernard Clark Campbell (Chicago: Aldine Publishing Company, 1972), 136–79.
    Note: In Trivers's chapter on reproductive investment and sexual selection, females do choose, but not according to aesthetic criteria.

one mating encounter; being "choosy" might actually be detrimental, as it would decrease the sheer number of copulations in which they participated. As most animals exhibited some form of differential parental investment, most animals probably participated in some form of female choice and male-male competition over mates. Trivers's point was that sexual selection was far more common than biologists had realized.

Citation analysis of Trivers's papers from the early 1970s bears out his claim that he remained beneath public notice until the publication of Wilson's *Sociobiology*.[24] However, it is unlikely that Wilson's book was solely responsible for Trivers's newfound fame. Trivers's presence at Harvard (and resultant contact with DeVore, Mayr, and Wilson) and the publication of his papers were remarkably well-timed to spread word of his results. When primatologist Bernard Campbell had approached Mayr about submitting a paper on sexual selection for the *Descent of Man* centennial volume, Mayr not only agreed to participate, but also passed along Trivers's name as another good candidate. Besides his own writing, Trivers also helped Wilson edit and revise *Sociobiology* before Wilson submitted the final draft of the manuscript to the publisher, and Wilson included many of Trivers's ideas in his chapters on sex and human behavior.[25] Further, as one of Mayr's honorary graduate students, Trivers was invited to attend Mayr's 1974 conferences on the evolutionary synthesis (with the added stipulation that he keep his mouth shut).[26] The community of biologists interested in evolution and behavior in the 1970s was growing rapidly. Trivers was in an excellent position in the early 1970s to publish his work in highly visible places; within a few years, organismal biologists on both sides of the Atlantic had read and cited his work.[27]

The centrality of Trivers's work to subsequent field research on sexual selection cannot be attributed to any single cause; he had good timing and a strict individual-based adaptationist framework, and evolutionary biologists generally were paying increasing attention to questions of human social and sexual behavior. Additionally, Trivers did not produce any significant papers on the topic in subsequent decades; if biologists wanted to cite a Trivers paper on sexual selection, of necessity they returned to his 1972 chapter. Yet perhaps the most important reason for Trivers's success was that his chapter provided convenient one-stop shopping; he synthesized the complex mathematical arguments of John Maynard Smith and William D. Hamilton with the experimental evidence of population geneticists such as Claudine Petit and Lee Ehrman, illustrating how they could be applied to field research, in a short, easy-to-read form.

In 1964, Hamilton had published two papers that dramatically changed the face of organismal biology in subsequent decades.[28] He argued that an individual's evolutionary fitness should not be defined solely by the number of offspring it produced, but should instead be conceived of as the individual's total genetic contribution to the next generation, which included the reproductive success of its siblings. This explained a phenomenon he called "kin selection." Imagine that my brother and I both produce two children. According to kin selection, my genetic contribution to the next generation would be calculated as follows. Each of my children would possess genes inherited from me (50%) and from my partner (50%). Additionally, because my brother and I shared 50% of our genetic makeup through our parents, each of his children (each possessing 50% of his genes) would, on average, share 25% of my genetic makeup. My total genetic contribution to the next generation would therefore be 0.5 (my first child) + 0.5 (my second child) + 0.25 (my brother's first child) + 0.25 (my brother's second child), or 1.5. So, rather than merely maintaining my genes in the population (1.0), I would have increased the percentage of my genes in the next generation, assuming that the number of individuals in the population remained constant. Hamilton's theory of kin selection thus explained why an individual would risk her own neck to help her family members: apparently altruistic behaviors were really self-serving in terms of evolutionary fitness. Trivers applied Hamiltonian kin selection to theories of female choice and suggested that females might be under evolutionary pressure to ensure that a single male sired all her offspring in a season's brood. If her offspring were highly related to one another (sharing 50% of their genetic makeup if sired by the same father, as opposed to only 25% if multiple fathers), that would tend to decrease competition among them and thus increase their chances of survival. Females should therefore be highly choosy in selecting their sole partner each breeding season. For males who didn't participate in childcare, however, it still made evolutionary sense to be as promiscuous as possible.[29]

George C. Williams's *Adaptation and Natural Selection*, published in 1966, also provided an important theoretical stimulus for Trivers's "Parental Investment and Sexual Selection."[30] By then, one of the last bastions of the notion of group-level selection was the process of mating and raising young, the successful union of a male and female in a cooperative family unit.[31] Williams's book served as a focusing lens for biologists' critiques of group selection by advocating a version of natural selection based solely on the reproductive fitness of individuals.[32] When it came to reproduction, Williams cited Bateman's paper on reproductive investment and contended that "the greater promiscuity of the male and greater caution and discrimination of the female is found in animals generally."[33] In other words, male and female reproductive interests did not necessarily accord and, over time, natural selection had acted to make males more aggressive and territorial than females. Yet, for Williams, there could be no necessary distinction between natural and sexual selection, as each contributed to the overall number of offspring left by an individual.[34]

Building on such theoretical framing, and inspired by Trivers's 1972 chapter, other biologists also began to offer hypotheses about how female choice might act in nature. What is most remarkable about these discussions is that nobody questioned the ability of females to choose their mates, only the means by which such choices might occur. Amos Zahavi, an Israeli ornithologist, suggested that females might choose brightly colored males because those males were of a higher quality (were more evolutionarily fit in an absolute sense) than duller males.[35] Zahavi argued that this proposition solved two problems in the evolution of social behavior. First, female preference for a trait did not need to be linked to male expression of that trait, as Fisher's theory of runaway sexual selection presupposed. Second, if females preferred to mate with a male exhibiting an apparently altruistic trait, then the appearance of an altruistic trait could be due to sexual selection acting on individuals. Although it appeared that females were choosing a trait that decreased males' chances of survival (sexual selection thereby acting in opposition to natural selection), in actuality, only the males that were extraordinarily fit with regard to all other traits could afford to exhibit such a handicap. Female preference, therefore, was evolutionarily adaptive because it allowed the female to judge the overall quality of her mate and perhaps even increased group cohesion, all in the name of individual selection. A few years later, Hamilton and Marlene Zuk, then one of his graduate students at the University of Michigan, suggested that females might prefer bright males because bright males were more resistant to parasites than their duller competitors.[36] In both cases, female choice was transformed from an arbitrary preference into a true judgment of quality—sexual selection need not act in opposition to natural selection but could instead act in conjunction with natural selection, ensuring that high-quality individuals were more likely to both survive and produce more offspring.

By 1975, after nearly four decades, these scientific discussions of sexual selection also reentered the *public* limelight as part of a brewing controversy over the biological basis of human behavior that followed publication of entomologist Edward O. Wilson's *Sociobiology: A New Synthesis*. In *Sociobiology*, Wilson applied biological theories of behavior in animals to understanding patterns of social and sexual behavior in humans. Although his application of biological theory to human nature had a long tradition, Wilson perhaps took his argument even farther. He argued that "scientists and humanists should consider together the possibility that the time has come for ethics to be removed temporarily from the hands of the philosopher and biologicized."[37] His claim that the very morality of human culture should be analyzed as a biological trait engendered considerable consternation among both humanists and scientists. The resulting debate over the validity of Wilson's approach erased much of the previous tradition of using evolutionary theory to understand human behavior.[38] In 1980, for example, evolutionary biologist John Alcock wrote to Mayr lamenting this forgotten history: "in many ways I wish Wilson had given his book the title, Evolution and Social Behavior, or something equally unsnappy. As it is, there seems to be a widespread notion that he invented an entirely new discipline."[39] The ensuing debate over Wilson's book, among academics and non-academics alike, popularized organismal biologists' use of biological theories to understand human behavior. Wilson applied evolutionary theory to human behavior in *Sociobiology* to counter what he saw as an alarming tendency in the social sciences to think of human behavior solely in cultural terms. Trivers, in his characteristically direct fashion, is fond of using the phrase "the social so-called sciences" to make much the same point.[40] It also seems likely that Wilson's emphasis on human behavior in *Sociobiology* acted as a corrective to the public impression that the future of biological research rested solely in molecular biology.

In his chapter "Sex and Society," Wilson touted sex itself as "an antisocial force in evolution. Bonds are formed between individuals in spite of sex and not because of it."[41] He contended that differences between male and female sexual behavior were biological and therefore not the result of selection acting differently on men and

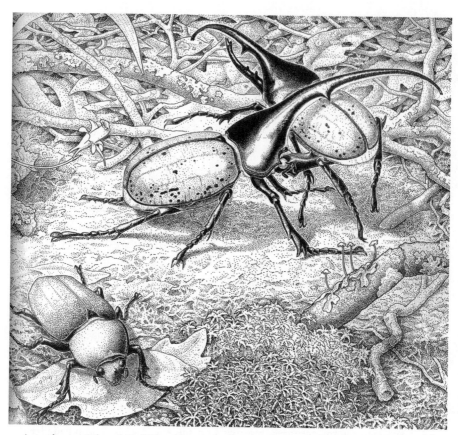

According to Edward O. Wilson's version of Darwinian sexual selection, females are sexually coy and extremely selective about their prospective mates. Here, a female Hercules beetle (foreground) awaits the attentions of whichever male is victorious in battle (background). Original drawing by Sarah Landry, reprinted by permission of the publisher from *Sociobiology: The New Synthesis* by Edward O. Wilson, p. 323, Cambridge, Mass.: The Belknap Press of Harvard University Press, copyright © 1975, 2000 by the President and Fellows of Harvard College.

women. Sex differences were not cultural, but genetic. Males, Wilson posited, were salesmen, and females had evolved to resist their efforts. In other words, by playing hard to get, females could land themselves a better (more vigorous) mate.

When Richards Dawkins's *The Selfish Gene* hit bookstore shelves a year after *Sociobiology*, it was a far slimmer volume and written in Dawkins's characteristic attention-grabbing style. It sold brilliantly.[42] In the preface, he professed: "We are survival machines—robot vehicles blindly programmed to preserve the selfish molecules know as genes." Yet Dawkins also noted carefully that he was "not advocating a morality based on evolution."[43] Trained as an ethologist at Oxford University under Niko Tinbergen's guidance, Dawkins saw his book as an ethological treatise for the layperson, infused with the new advances in theoretical biology from George C. Williams, John Maynard Smith, William D. Hamilton, and Robert Trivers.[44] Trivers says he was delighted when Dawkins asked him to write a foreword for the book, and then crushed when the second edition appeared in 1989 without his foreword, which Trivers attributed to their disintegrating friendship.[45]

In *The Selfish Gene*, Dawkins explicitly adopted several of Trivers's ideas concerning evolution and sex. Perhaps most notably, he incorporated Trivers's analysis of differential parental investment as the basis for sex-specific behaviors in animals and humans. Dawkins suggested that all behavioral differences between the sexes had a single root cause: sex differences in the size (and therefore energetic cost) of gametes. He argued that the significantly smaller size and greater quantity of male sperm compared with female eggs enabled males to reproduce at a virtually unlimited rate, while human females, for example, typically bore only one child at a time.[46] Such universal choosiness, however, did not mean that all females used the same rubric to select a mate; some might look for a male with good parental or provisioning skills, others for a male with the highest-quality genes. In contrast, males were consistently more profligate in their sexual attentions. By the second edition of *The Selfish Gene*, Dawkins had softened this claim. Instead of arguing that the relative size of gametes *caused* behavioral differences in adults, he suggested that both gamete size and adult behavior were *effects* of having two sexes to allow for genetic recombination—a resorting of genes from one generation to the next.[47] Like Darwin before him, Dawkins thought that if a biologist were to analyze human sexual systems based on appearance alone, he (or she) would be likely to think that in humans, the process of sexual selection was reversed, because it was the females of the species that spent so much time and attention on their appearance, while the males held the power of mate selection.[48] However, Dawkins was joking and his meaning remained clear—that sexual selection was not truly reversed in humans, despite outward appearances. Dawkins adapted the rationalist language of game theory as shorthand for long-term selection acting on animal instincts.[49]

Such vehement assertions of the biological basis of male sexual aggressiveness and female sexual reticence seemed designed to quell claims of contemporary feminists that women deserved as much sexual and economic freedom as men. The rise of second-wave feminism in the 1960s and the sexual revolution drew popular attention to women's sexual choices in human society. In her 1963 bestseller *The Feminine Mystique*, for example, Betty Friedan urged suburban housewives to recognize their frustrations with making beds, shopping for groceries, matching slipcover material, and chauffeuring the children.[50] Women needed to break down the stereotypes that defined femininity, she claimed, and choose how they wanted to live their lives. Published in 1966, the psychological studies of William Masters and Virginia Johnson simultaneously illuminated the psychology and physiology of the female sexual response, including the capacity of women for multiple orgasms. For women who could afford the oral contraceptives newly approved by the U.S. Food and Drug Administration, the choice to reproduce could be regulated through modern medicine in the form of a pill.[51] The heightened awareness of women's social roles and sexual lives contributed substantially to the enormous media attention generated by the publication of *Sociobiology* and *The Selfish Gene*. As in the late nineteenth century, although some women embraced the renewed biological equation of femininity with reproduction, others strongly objected.

Following the publication of both books, many biologists and feminists resisted sociobiological characterizations of men as sexually vigorous and women as passive or coy.[52] These reductionist stereotypes, feminists argued, simply reinscribed postwar gender roles in terms of biology or, even worse, hearkened back to Victorian conceptions of proper behavior for men and women.[53]

These highly visible debates on the validity of sociobiology's claims about the biological basis of human sexual behavior also helped to erase the history of female choice and sexual selection in the preceding decades. A young feminist anthropologist, Sarah Blaffer Hrdy, was one of the few people to read Wilson's "Sex and Society," before the publication of *Sociobiology*; at the time, Hrdy was a Ph.D. student in anthropology at Harvard, where Wilson worked. Another person who probably read the chapter was Trivers.[54] Whether or not Hrdy's objections to Wilson's book coalesced the first time she read his manuscript, in 1986 Hrdy published a scathing critique of the enduring role of Bateman's 1948 "coy" female and "active" male in contemporary evolutionary theory. In this critique, she identified sexual selection as "one of the crown jewels of the Darwinian approach basic to sociobiology" and argued that the theory of sexual selection was based on "partially true assumptions" that stereotyped females as sexually passive, yet highly discriminating, and males as sexually aggressive and undiscriminating.[55] Hrdy contended that Trivers had uncritically imported Bateman's sexual stereotypes into his work. As Trivers's 1972 essay on parental investment gained in popularity, his readers also incorporated the myth of the coy female into their own biological theorizing about the sexual relations between males and females.[56]

In her article, Hrdy suggested that such a model gained acceptance within the biological community because of scientific bias and that the recent incorporation of women into the biological sciences had begun to alter the application of sexist stereotypical human behavior to animals.[57] She chose her target wisely. By the mid-1980s, when Hrdy published her critique of the "coy female," Trivers's chapter was being cited an average of 130 times per year (this number has only grown since then), and his central role in the rebirth narrative of sexual selection was well-secured.[58] Hrdy's criticism of Trivers's and Bateman's coy female was intended as a general critique of experimental practice and theories of sexual selection research, and it serves to highlight how crucial the ethological model of mate choice had become to the animal behavior community.

When Trivers adopted Bateman's description of a coy female as a universal ideal, he united it with the population geneticists' interest in understanding the process of evolution. The new enthusiasm for sexual selection research that developed in the 1970s grew out of ethological, theoretical, and population genetics approaches to behavior and evolution, but merged them in such a way as to make sexual selection's historical past difficult to parse. Female choice formed the crux of this surge in interest, as Trivers used differential parental investment once again to provide a theoretical grounding for his seemingly innovative claim that females could and should be picky in their choice of mates.[59]

As a result of the scientific and feminist outcry over sociobiology and its associated ideas, coupled with the explosion of interest in field research on animal behavior generally, it seemed that sexual selection and female choice had both been reborn in the 1970s through the work of men like Trivers, Hamilton, Williams, and Maynard Smith. Field biologists hailed Trivers's "Parental Investment and Sexual Selection," in particular, as the scientific spark of the sexual selection revolution, popularized (with all the connotations typically embedded in this word) by *Sociobiology* and *The Selfish Gene*. For the first time since the 1930s, biologists framed female choice primarily as a mechanism to explain sexual difference and to be investigated by organismal biologists in the field. Female choice once again became a process by which a discriminating female selected the most stimulating male with whom to mate, as well as a mechanism that explained morphological and behavioral differences between the sexes.

Constructing the History of Sexual Selection

The attention of many organismal biologists with an eye for history was captured by the highly publicized Darwin centennial celebrations of 1959 and the attendant publications.[60] At the University of Chicago, for example, the Darwin Centennial Celebration lasted for four days and included a musical about Darwin, a talk by his grandson, and a convocation delivered by Julian Huxley. In addition to these highly visible public events, the intellectual work of the celebration included a care-fully orchestrated and recorded series of panel discussions: on the origin of life, the evolution of life, man as an organism, the evolution of mind, and social and cultural evolution. Although these last two panels did not deal exclusively with human evolution, they constructed human behavior and culture as an explanatory goal for research on the mind and behavior of animals. The papers presented by the scientists in these sessions were published in three volumes edited by the Univer-sity of Chicago anthropologist Sol Tax.[61] Similar celebrations were held in Britain, but there, publications were more historically oriented. Evolutionary embryologist Gavin de Beer edited and published the first transcription of Darwin's notebooks on the transmutation of species in 1960. These published notebooks made Darwin's thoughts accessible to biologists and other nonhistorians. An American comparative anatomist, Michael Ghiselin, was so inspired by reading Darwin's notebooks that he read all of Darwin's published works as well and went on to write an award-winning historical volume, *The Triumph of the Darwinian Method.*[62] Mayr, too, caught the historical bug. In 1974, he used his position as a visiting professor at the University of Minnesota as an opportunity to develop a course on the history of biology, and he organized two conferences designed to correct what he identified as bias toward the molecular in the history of twentieth-century biology.[63]

Although molecular biology as a methodological approach originated in the 1930s, it was not until the 1960s that biologists began to respond to its pervasive in-fluence by restructuring the traditional subject-based disciplinary boundaries within biology.[64] Efforts to reform and reorganize the structure of biological fields of study originated both among molecular biologists seeking to find disciplinary traction and among biologists who thought that their fields of study were being pushed out of the limelight by molecular biology. The battle of "organismal" biologists, as they called themselves, against molecular biology came from several fronts. Organismal biologists sought to create an identity for their research approach as philosophically separate from, yet equal in status to, molecular biology. Other biologists adopted Mayr's distinction between proximate causes (immediate physiological or mechani-cal causes of traits) and ultimate causes (longer-term historical and functional causes of traits) as a way of parsing the different research agendas that characterized mo-lecular and organismal biology.[65]

In addition, organismal biologists began to speak to subjects historically ad-dressed by the human social sciences. By asserting the biological basis of human be-havior, they ensured, for at least the foreseeable future, that the questions addressed by molecular biologists could not encroach further into their disciplinary territory. Perhaps because of the increased fracturing of biology as a discipline, organismal biologists also sought to maintain biology as a unified subject, with evolutionary theory at its core. (Evolutionary theory, of course, they identified as part of their own jurisdiction.) Given the attempt to define evolution as the unifying factor of modern biology, debates over the organizational level at which selection acted and biological reproduction occurred became particularly contentious and centered on gene-level versus organism-level selection. Equally important were the ways in which organismal biologists used histories of the evolutionary synthesis as philosophical tools defining their vision of "true" biology.[66]

The history of sexual selection and female choice became entangled within these disciplinary divides at a moment when organismal biologists were looking for ways to claim authority in a molecular age. Biologists and historians of biology came to identify sexual selection as an "organismal" field of inquiry that had undergone a period of eclipse because the major advocates of evolutionary biology in the twentieth century (Dobzhansky, Simpson, Huxley, and especially Mayr) failed to understand the importance of female choice during the heyday of their careers. According to this narrative, it was not until the 1970s that organismal biologists began to pay attention to sexual selection, thanks to biologists' figurative stampede to Hamiltonian kin selection, games of rational choice, and Robert Trivers.[67] Yet negotiations over the proper history of sexual selection started even before Trivers published his 1972 essay. Following standard rhetorical modes in the sciences, short histories of sexual selection often appeared in the introductions to scientific monographs or in the first few paragraphs of a review paper. Although all identified the same few key figures—Charles Darwin, Ronald Fisher, and Julian Huxley—these histories varied considerably in their interpretation of the roles played by these men.[68] For example, the histories generally agreed that Huxley's 1938 papers were crucial to some biologists' dismissal of mate choice as a cause of sexual dimorphism, but they disagreed on who found Huxley's theories persuasive and how long the eclipse of interest lasted. These histories also differed considerably in their treatment of population genetics. Whereas Mayr tended to blame the "fathers" of population genetics for the disappearance of mate choice, contemporary population geneticists emphasized the importance of their research in maintaining interest in mate choice as an evolutionary mechanism during the so-called eclipse. By the mid-1980s, sexual selection and female choice fell clearly within the jurisdiction of organismal biology, and historical accounts of sexual selection stabilized.

In what became the standard story of sexual selection, the lack of scholarly interest in female choice lasted for almost a century, from Alfred Russel Wallace's dismissal of it as a viable evolutionary mechanism by 1877 to its "resurrection" by Trivers in 1972.[69] Who was to blame? Emphases varied, but the cast of characters did not. Wallace was a victim of his Victorian sensibilities. Huxley was a bad naturalist or was swayed by his commitment to monogamy as a paradigmatic mating system in the natural world. Geneticists such as Fisher, Haldane, and Wright insisted that evolution acted on genes, not individuals, redefining natural selection as genetic contribution to the next generation and thereby obviating the need for sexual selection as theoretically distinct from natural selection. Of these men, Fisher's role in the standard story remained the most ambiguous, as some histories, mostly written by population geneticists, portrayed him as a neglected genius whose theory of runaway sexual selection failed to catch biologists' attentions until the 1970s. Huxley was the primary scapegoat of these histories, with Wallace playing a secondary role. Huxley was a politically easy target. He had stopped active research during the Second World War, and in the United States was seen more as an icon of the evolutionary synthesis than as a practicing scientist.[70] By way of contrast, Mayr, Dobzhansky, and Simpson, if not actually revered, were at least sufficiently active within the evolutionary community to have strong allies who had the wherewithal to strike back if maligned.[71]

Robert K. Selander wrote some of the first retrospective reviews of the history of sexual selection, starting with his 1965 paper "On Mating Systems and Sexual Selection." An ornithologist by training, Selander later turned to research on protein polymorphism and the population genetics of natural populations. In his article, he argued that both Fisher and Huxley recognized and pointed to the difference between the "interspecific" mate choice needed to identify a mate of the correct species and the "intraspecific" mate choice among multiple potential mates of the same species. After Huxley, Selander claimed, researchers chose to investigate only the former aspect of mate choice and ignored intraspecific choice. He did not attempt to explain this lack of interest, merely stating that the reasons were "not entirely apparent." Just seven years later, for Bernard Campbell's *Sexual Selection and the Descent of Man, 1871–1971,* Selander wrote a second review of mating systems in birds. In his concluding comment, he described the "synthetic theory of evolution" as breaking the "mutationists' hold on evolutionary biology" that had started in the 1930s. It was only in recent years, he continued, that "the study of social systems has demonstrated the immense power of natural selection (including sexual selection) and the complexity and subtlety of adaptations it can generate."[72]

Selander's brief historical synopses illustrate three notable characteristics of what became the standard story. First, early twentieth-century theories about female choice in humans did not appear in his narrative. Second, he established sexual selection as an organismal science by blaming the laboratory sciences, and genetics in particular, for its underappreciation. Third, he dismissed the research of ethologists and experimental population geneticists on interspecific female choice to maintain reproductive isolation between species as being outside sexual selection's more important concern with intraspecific female choice. Only some kinds of female choice counted.

Thanks to the success of Trivers's 1972 paper in the growing field of animal behavior studies, Campbell's edited volume—and thus Selander's review—became part of a new historical authority, wiping out a more interdisciplinary picture of the history of female choice. However, Campbell's volume contained six other chapters also devoted entirely to sexual selection—a large and perhaps unlikely assemblage for a theory in such supposed disrepute.[73] Lee Ehrman, for example, contributed a chapter on genetics and sexual selection in *Drosophila*. Dobzhansky wrote on genetics and the races of man, and Mayr offered a theoretical comparison of sexual selection and natural selection.[74] Yet biologists do not cite these chapters with anywhere near the frequency with which they cite the Trivers chapter. It is tempting to attribute biologists' later lack of attention to these other contributions to the chapters' emphasis on female choice as a mechanism for reproductive isolation, rather than sexual dimorphism; yet clearly, more must be at stake.[75]

It took more than a decade for the eclipse narrative of sexual selection to stabilize, and reviews of Campbell's volume emphasized the variety of definitions used by the authors when discussing both female choice and sexual selection.[76] Whereas Selander referred to sexual selection as a process causing "display characters or weapons . . . directly involved in obtaining mates and copulations," Lee Ehrman defined sexual selection as all mechanisms causing nonrandom mating in a population, including assortative mating (when, for example, large individuals tended to pair with large mates, and small with small). Ernest W. Caspari specifically excluded assortative mating from his definition of sexual selection. To further confuse the point, Mayr used his chapter as an opportunity to contrast natural and sexual selection; true fitness, the result of natural selection, he associated with the vigor of individuals, while sexual selection explained only the "mere reproductive advantage" of males and females.[77] Here, Mayr used his definition of sexual selection as way of downplaying the importance of reproductive fitness as a mechanism of evolution (counter to Williams and Hamilton), a point to which we will soon return. The Campbell volume heralded the dawn of a new era in sexual selection research, yet it is also clear that in 1972, a consensus definition of sexual selection was still up for grabs, as were the disciplinary affiliations of biologists interested in the theory.

In retrospect, one review of the Campbell volume stands out from the rest: primatologist Adrian Zihlman chose to emphasize the human dimension of the eclipse narrative of sexual selection. She contended that the eclipse of biologists' interest in sexual selection centered not on female choice in animals generally, but specifically on how female choice as an evolutionary mechanism was thought to act in humans.[78] Biologists' interest in the role of female choice in humans had been largely ignored from the 1930s to the 1960s, Zihlman continued. In response to Darwin's claim that mate choice was reversed in humans (because males, rather than females, chose their mates), she suggested that he had been "limited by the cultural milieu of Victorian England, where women were seen as passive and uninitiating."[79] Zihlman lauded Campbell's volume as an attempt to distance sexual selection theory from Darwin's dichotomous view of mate choice and to recognize the importance of female choice in primate and human evolution as well. Zihlman's emphasis on the human context as important for understanding the history of sexual selection was absolutely right, but her thoughts appeared in the form of a book review and remained uncited by her peers.

The "standard" history received another boost with the publication of Peter O'Donald's *Genetic Models of Sexual Selection* in 1980. In this book, O'Donald translated Ronald Fisher's runaway theory of sexual selection from prose into mathematical terms and developed his own population genetics mathematical models of sexual selection theory.[80] He also included a review chapter that pointed to Huxley's central role in causing a decline of interest in sexual selection. O'Donald continued by dismissing Huxley's analysis as "hopelessly confused" because he had conflated sexual selection with natural selection, and natural selection with group selection. Given the confusion that O'Donald thought was so evident in Huxley's prose, he could not imagine how Huxley had been so successful in repressing other biologists' interest in sexual selection.

O'Donald's narrative differed from Selander's version of the history of sexual selection in one key aspect: he credited the Campbell volume with *completing* the reinstatement of Darwin's second theory, not sparking it. As a mathematical population geneticist who began his research on the mathematics of sexual selection in the mid-1960s, O'Donald's interest in sexual selection had been piqued by the 1958 republication of Fisher's *Genetical Theory of Natural Selection*.[81] O'Donald's connection to the British ecological and ethological communities and his familiarity with *Drosophila* research led him to cite the work of Lee Ehrman, Claudine Petit, and Eliot Spiess on female mate choice in fruit flies. For O'Donald, the surge of interest in sexual selection started a decade earlier than 1972, with the work of these population geneticists.

In reviews of O'Donald's *Genetical Theory*, other biologists interested in animal behavior commented on his historical account. Russ Lande, who had recently

earned his Ph.D. in biology from Harvard (and was an assistant professor of theoretical biology at the University of Chicago), emphasized the crucial role of Huxley's
1938 papers in causing the eclipse as a result of Huxley's inability to distinguish
between individual and group selection. "Until recently," Lande suggested, "it was
not realized by many workers that under sexual selection the fitness of a trait with
respect to mating success can override its value for survival, decreasing the average
total fitness in a population."[82] The idea that the key figures involved in the synthesis
found sexual selection to be a less than engaging topic was borne out by the con-
tents of Ernst Mayr and William Provine's 1980 volume *The Evolutionary Synthesis*.[83]
Sexual selection was barely mentioned, and they further suggested that behavior and
evolution had not been synthesized until the 1970s. Thus, for Lande, recent interest
in sexual selection was caused by a renewed appreciation that sexual selection could
act in opposition to natural selection.

Yet, when population geneticist Eliot Spiess reviewed O'Donald's book, he emphasized the primacy of his and his colleagues' research in population genetics during the 1950s and 1960s as the cause of the recent vitality of sexual selection within
evolutionary biology. Spiess also referred to Huxley as the cause of the eclipse, but
noted sexual selection's "reinstatement after 1950 with experimental verification and
culmination in the publication of the symposium papers (edited by B. Campbell)
celebrating the centennial of Darwin's publication."[84] French geneticist Claudine
Petit remembers her own intellectual history in similar terms. She had taken Huxley's papers as a challenge for further research. Rather than dampening her interest
in sexual selection, Huxley's claims inspired her.[85] In this version of the history,
sexual selection was eclipsed only between 1938 and 1950! If this were true, it should
hardly come as a surprise, given that many biologists of this generation interrupted
their research during the Second World War.

In a later, more comprehensive review of female choice in fruit flies, Spiess again
stressed the importance of *Drosophila* research in the postwar period as an influence
on sexual selection research in the 1970s. He argued that he, like many biologists
educated in the 1940s, "had the impression that Darwin's concept of female choice
of mates in the evolution of secondary sexual characters was to be considered a form
of natural selection." Population geneticists had overlooked Huxley's "shortsighted
criticisms" of Darwinian sexual selection, and they "began to analyze genetic variation as it affects reproductive success of both sexes, particularly in *Drosophila*."[86]

Spiess carefully distinguished between two potential meanings of the phrase
"mate choice" that were often confused in the evolutionary literature. Spiess (much
like Huxley in 1938) hoped that the single term might be replaced by two different
terms, to clarify biologists' intentions. Spiess proposed using *choice* to "refer to the

selection by an individual from a set of alternatives" and *preference* to refer to "a bias for or against a particular choice from those alternatives." In his nomenclatural scheme, female preference implied a stimulatory or threshold model of mate selection much like that proposed by ethologists in the 1950s and 1960s. Female choice, on the other hand, implied mental comparisons of some sort on the part of the animal. Nomenclatural confusion over these same two models of female choice had inspired Huxley's attempts to distinguish sexual selection from natural selection in the 1930s, yet his papers clearly failed to solve the problem for most biologists, as theoretical disputes were similarly current through the 1980s.

In 1983, Patrick Bateson published a collection of papers in the volume *Mate Choice*, taking care to avoid confusing the concepts of mate choice and sexual selection.[87] A midcareer ethologist, Bateson had centered his research interests on reproductive behavior and imprinting in birds. His volume originated as a conference in 1981, after which the contributing authors transformed their research talks into review articles for the book. In the introduction, Bateson implied that most biologists would be tempted to assume that all mate choice should count as sexual selection, and he admonished them to remember that some forms of mate choice simply identified a mate of the correct species. Bateson's concerns in 1983 echoed Huxley's in 1938, yet they were embedded in a radically different disciplinary context. In the first decades of the twentieth century, zoologists tended to assume that female choice occurred within a single species. Huxley had sought to correct this view by insisting that in the animal world, the evolution of courtship displays was driven less by a female's choice of one male among those of her own species (sexual selection) than by the need for a female to choose a male of her own species (natural selection). Within their new research program, evolutionary synthesists of the 1940s and 1950s successfully shifted zoological research in just this way. Half a century later, Bateson's primary audience consisted of young biologists interested in animal behavior who sought to distinguish themselves from these classical evolutionary theorists by interpreting all evidence of mate choice as evidence of sexual selection. How times had changed!

In general, young professors of animal behavior such as Bateson laid the blame for the neglect of sexual selection as a research topic at the feet of all the custodians of the evolutionary synthesis, not just Julian Huxley. In a 1984 review of Bateson's book, entitled "Revenge of the Ugly Duckling," Mark Kirkpatrick (who had just received his Ph.D. in zoology from the University of Washington) highlighted the lack of interest during the period of the modern synthesis in the question of mate choice as a mechanism for morphological and behavioral differences between the sexes. "Only in the last two decades," Kirkpatrick argued, "have mate choice and sexual selection received broad attention from evolutionary biologists. With the notable exception of contributions from E. B. Poulton and R. A. Fisher, very little was added to Darwin's ideas on the subject for nearly a century." Kirkpatrick suggested that *all* the major evolutionary figures, including Huxley, Mayr, Dobzhansky, David Lack, and Bernard Rensch (all of whom Mayr and Provine included in their 1980 book *The Evolutionary Synthesis*), "gave very little weight to sexual selection," not out of "mere oversight," but because they sought to demonstrate that mate choice was a "side effect of selection for interspecific isolating mechanisms." Only recently had biologists begun to question the wisdom of their elders' conclusions: "mate choice is now being considered as an important evolutionary agent in its own right. Evolutionary biology's ugly duckling is undergoing a metamorphosis."[88] Kirkpatrick's review effectively accused Huxley and Mayr of short-sighted inattention to sexual selection. Mayr, in particular, should have known better, Kirkpatrick implied, because he was an authority on birds of paradise and, by the 1980s, birds of paradise were one of the textbook examples of sexual selection in birds.

Not surprisingly, Ernst Mayr, guardian of all synthesis-related issues, urgently felt the need to defend himself and those he considered his closest colleagues. In response to the review, Mayr sent Kirkpatrick a letter in which he agreed that he had not emphasized sexual selection in his 1942 book. By way of explanation, Mayr posited that "a neglect of sexual selection was perhaps not excusable, but certainly understandable." He provided several reasons for his neglect of sexual selection. Primarily, it was Huxley's fault—Huxley was the animal behavior expert and everybody believed him. In addition, Mayr postulated that exotic bird species without extended courtship prior to mating would tend to mate with individuals from the wrong species at a greater frequency than would those with extended courtship. He had therefore concluded that the extravagant plumage of male birds of paradise, and correlated female choice, existed to ensure proper species identification, not because of sexual selection. As a final cause, and perhaps most interestingly, Mayr added that the redefinition of fitness as a generational change in gene frequency perpetrated by mathematical population geneticists made any distinction between natural selection and sexual selection "irrelevant."[89] Mayr's statement in his letter to Kirkpatrick came straight from his summary of the reasons for the neglect of sexual selection presented in his 1982 book, *The Growth of Biological Thought*.[90] In his book, Mayr added that it was only when individuals were reinstated as the target of selection (rather than genes) that it became "respectable" to discuss sexual selection once more. With the obvious exception of Huxley (whose reputation he threw under the bus), Mayr swiftly passed the blame for the neglect of sexual selection from the shoulders of the guardians of the modern synthesis into the laps of the mathematical population geneticists.

Although all historical accounts emphasized Julian Huxley's role in suppressing interest in sexual selection, the differences between the accounts are quite revealing. Scientists from at least three biological communities were involved in creating the history of sexual selection: population geneticists, theoretical mathematical geneticists, and zoologists. Population geneticists, such as Spiess and Petit, worked primarily on fruit flies and insisted that the eclipse of sexual selection lasted only until the mid-1950s. Theoretical mathematical geneticists, as exemplified by O'Donald and Maynard Smith, illustrated the eclipse of sexual selection by emphasizing that Fisher's theory of runaway sexual selection had been ignored by zoologists—probably, they implied, because Fisher's mathematics was too difficult for naturalists to understand.[91] Yet according to zoologists, including Mayr and Selander, it was the early population geneticists who were primarily to blame for the eclipse of sexual selection. They insisted that geneticists invented a definition of natural selection (reproductive contribution to the next generation) that obviated the need for sexual selection as a distinct concept and, as a result, scientific interest in sexual selection vanished.

Generational training also made a difference, as many of the younger faculty members sought to establish the study of sexual selection as a new and revolutionary field of study. Most biologists who worked on sexual selection in the 1980s came from a new generation of academics who distanced themselves from older evolutionary biologists by blaming them for the neglect of the theory. These scholars also ignored laboratory geneticists' claims that they had been working on female choice throughout the 1950s and 1960s.

A quick glance through the pages of subsequent editions of John Alcock's *Animal Behavior* (which for years has been a standard textbook in American undergraduate classes on the topic) illustrates the slow instantiation of the eclipse narrative among evolutionary biologists. In the first edition of the textbook, published in 1975, neither Trivers's name nor the subject of sexual selection was mentioned. The second edition, published four years later, described Trivers as having "proposed" that sexual selection occurred because the sexes differed in their degree of investment in reproduction and childcare. By 1984, Trivers's stock, and that of sexual selection, had risen considerably; in the third edition of *Animal Behavior*, Trivers "invented" the concept of parental investment, and Alcock devoted two entire chapters to the subject of sexual selection. It was not until the fourth edition (1989), however, that Alcock mentioned the importance of Bateman's contribution to Trivers's ideas, discussed Fisher's concept of runaway sexual selection, and made use of the eclipse history of sexual selection. "After Darwin," Alcock wrote, "the concept of sexual selection was largely ignored until 1972, when Robert L. Trivers resurrected the topic by drawing attention to a study by A. J. Bateman."[92] The importance of Trivers's research to the historical narrative of sexual selection was closely linked to the currency of the eclipse model of sexual selection, and neither was firmly secured until the later 1980s.[93]

By the late 1970s, sexual selection by female choice was a hot topic. Not only did everyone want a piece of the action, but they also wanted to claim that their field was not responsible for the eclipse of interest in sexual selection. The cessation of negotiations over the history of sexual selection reflects organismal biologists' ultimate success in claiming sexual selection as an organismal subject.

The return of sexual selection as an exciting research program within organismal biology must be attributed to a variety of factors. Organismal biologists promoted animals as behavioral models for understanding human behavior. Game theory models of choice-based behaviors provided a new, nonanthropomorphic framework with which to discuss the apparently rational choices of animals. And Robert Trivers's 1972 essay made the research of population geneticists and theoretical biologists accessible and convenient for organismal biologists. Additionally, second-wave feminism and the debates over sociobiology created sufficient frisson to attract further public and scholarly attention to the issue of female choice.

The eclipse history of sexual selection is an exemplary case of practicing biologists' fascination with the history of their discipline and reflects the disciplinary affiliations of sexual selection in the 1980s. As a result, the history of sexual selection as a mechanism to ensure reproductive isolation between species, or to maintain the genetic heterozygosity of populations, was lost. Only sexual selection as an evolutionary mechanism causing sexual dimorphism was included. These disciplinary affinities reflect the slow division of postwar biologists into at least three communities: experimental laboratory biologists, organismal field biologists, and theoretical mathematical biologists. The eclipse history of female choice valorized the contributions of organismal biologists to understanding the evolutionary process in nature (read: in the field), to the detriment of contributions by either laboratory or theoretical biologists. With repetition, the history of sexual selection proffered by biologists became increasingly programmatic and historically naive. One historical retrospective, written in 1998, went so far as insisting that sexual selection was "the only major scientific theory ever to have been accepted after a century of condemnation."[94] Such a historical perspective was far from inevitable and masks considerable negotiation.

Certainly, blame for any eclipse in the history of sexual selection cannot be laid solely at the feet of female choice. Most Victorian and Edwardian biologists rejected the notion of all choice in animals, as they believed that only humans possessed the necessary mental machinery to choose an aesthetically pleasing mate. Alfred Russel Wallace did not receive sexual selection frostily but, much like Ronald Fisher, simply believed it was a likely mechanism of evolution only in humans, not in animals. Julian Huxley was not deeply hostile toward sexual selection, despite his lack of later publications on the topic. Certainly, ethologically trained zoologists after the Second World War were more interested in exploring the effects of natural selection on behavior than in how any behavior, even mate choice, might contribute to the evolutionary process. Yet population geneticists during the same period were fascinated by questions of how mating behaviors, including mate choice, could contribute to the evolution of a population. The revival of sexual selection as an important topic within field biology was slow, spanned many decades, and owed much of its theoretical justification for behavior as a mechanism of evolutionary change to population genetics' analyses of the mating behavior of fruit flies, ethological models of stimulatory behavior, and mathematical models of evolutionary change. Although biologists were by no means fixated on the importance of sexual selection, neither was female choice "condemned" for almost a century.

To have been successfully promulgated by contemporary biologists, then, the eclipse narrative of sexual selection must have served a purpose in defining their biological practice. The longevity of this historical narrative derives at least partially from the power and cachet it lent to organismal biologists in their struggle with molecular biologists for institutional funding and public recognition. In this, organismal biologists have succeeded remarkably.

In the 1960s and 1970s, organismal biologists turned to animal models in their search for solutions to human social problems. This strategy had a long-standing tradition in twentieth-century biological sciences, from Fisher's eugenic politics in the 1920s to the debates between Theodosius Dobzhansky and H. J. Muller over the genetic diversity of populations in the 1950s.[95] However, the "hardening" of evolutionary biology into a rigid adaptationist framework in the 1960s gave new credibility to biologists who sought to understand the biological basis of human behavior through an evolutionary lens.[96] An adaptationist framework permeated most other evolutionary conversations as well, from the coevolution of insect-plant interactions to paleobiology and evolutionary ecology.[97]

The success of this zoomorphic agenda within organismal biology was double-edged. Zoologists had been fighting to increase the prominence and popular recognition of biological research involving whole animals, in the face of what they called "reductionist" molecular biology. By using the term "reductionist," organismal biologists hoped to point out that one could not understand all life by analyzing the component parts of organisms. In organismal lingo, then, *reductionism* meant methodological analysis at a subcellular level of biological organization. After the publication of *Sociobiology*, the label "vulgar reductionism" was flung back at organismal biologists interested in the biological basis of human behavior.[98] Such critics of the sociobiological approach intended their aspersions differently. They sought to highlight the way that Edward O. Wilson and other sociobiologists assumed that all animal and human behavior had a heritable genetic basis. Their criticisms were not directed at the methodological level of analysis (which remained organismal), but at sociobiologists' presumption that all naturally occurring behaviors had a genetic basis and could therefore be modified by natural selection, without analyzing the genetic basis of the behaviors in question.[99]

Despite such concerns, popular interest in the behavior of wild (and domestic) animals has only grown, thanks to the devoted consumers of the Discovery Channel, National Geographic films, Animal Planet, and even the slightly less dramatic pages of the *New York Times*. The moral lesson provided by their sensational tales of charismatic megafauna is clear: we must preserve the wild animals and environments of this planet or we may lose our ability to understand our own biological identity.

Conclusion

This historical exploration of female choice has focused on how scientific ideas about choice-based mating behaviors in animals developed within different communities of biologists over a span of a hundred years, from the work of Charles Darwin and Alfred Russel Wallace in the late nineteenth century to sociobiological research in the late twentieth century. The language with which professional biologists of this period discussed the mating behavior of animals depended on both their disciplinary training and their beliefs concerning the human equivalent of animal courtship behavior. The simple words *female choice* carried a variety of connotations, depending on the particular material and theoretical cultures in which biologists worked, and biologists' conceptions of animal mating behavior and human courtship behavior have remained tightly interwoven. Despite the close association of female choice with Darwinian sexual selection theory at the endpoints of my story, the fate of female choice was not the same as that of sexual selection. Biologists' interest in female choice as a mechanism of speciation or genetic diversity remained strong throughout the twentieth century, while they found much less gripping the topic of sexual selection as a mechanism to explain sex differences. At mid-century, when female choice was reframed in terms of female sensitivity to male stimulatory courtship behavior, the concept was stripped of its rationalist implications, and studied eagerly by population geneticists. As notions of animal mind changed, so did the acceptability of female choice and sexual selection as evolutionary mechanisms.

Throughout much of the twentieth century, common scientific sense seemed to dictate that animals could not make a choice based on rational or aesthetic criteria. Such choices were reserved for the mental capacity of humans. Scientists who transgressed this animal-human cognitive divide were alternately accused of anthropomorphism or zoomorphism. Yet, even the meaning of these terms changed. At the turn of the century, anthropomorphism was an arrow directed at biologists who described the behavior of animals in overly sentimentalist terms. Very early in the twentieth century, anthropomorphism began to carry a slightly different mes-sage—one implying that rationality lay behind an animal's actions. With the rise of behaviorism, comparative psychology, and even ethology, "professional" scientists assumed the least degree of animal cognitive processes necessary to explain the be-havior of their animal subjects. Today, we tend to think of anthropomorphism as the process of over-emotionalizing animal behavior—my cat is angry with me, my parakeet misses me when I am gone—but for most of the historical actors in this story, animals were always emotional beings; it was not the emotional but the cog-nitive capacities of animal minds that they questioned. In the 1960s, biologists re-identified another way of transgressing the animal-human boundary: zoomorphism now became an arrow flung at ethologically trained zoologists and sociobiologists who seemed to describe human behavior solely in the nonrational terms typically reserved for animals. If *Homo sapiens* was just another animal species, a naked ape, then one of the primary qualities that distinguished us from our beastly brethren, our ability to reflect, had been lost in analyses of the biological basis of human behavior.

Biologists' concerns over anthropomorphism or zoomorphism were crucial to the different models of female choice implicit in each of these biological communi-ties. For example, ethologists' conception of female choice as a threshold behavior fit into their model of how behavior operated in nature more generally. Ethologists argued that some males were simply more efficient than others at getting females in the mood for sex; they did not invoke cognitive choices to explain observed be-haviors in animals. In making this argument, ethologists saw themselves as circum-venting earlier criticisms that the study of animal behavior could be nothing more than amateurish anthropomorphism.[1] Population geneticists, however, remained far less worried about their professional status and invoked choice-based behaviors in female flies without fear of incurring critiques of their linguistic descriptions.

Both anthropomorphic and zoomorphic strategies have proved to be enduring ways, albeit controversial and historically contingent, of theorizing and analyzing the behavior of animals and humans. They remained powerful explanatory frame-works within animal behavior research throughout the twentieth century, in part because both strategies emphasized the power of natural behavior in animals to determine normative behavior in humans, changing as the categories of "natural" and "normative" themselves changed.[2]

The history of female choice also offers a unique vantage point from which to investigate historical claims about the unification of evolutionary theory in the mid twentieth century and to attend to the shifting landscapes of theoretical, laboratory, and field research on questions of evolution and behavior. I raise here the implications of the history of female choice for understanding how animal behavior research changed as a function of evolutionary theory.

Philosophers and historians of biology continue to hail the modern evolutionary synthesis as one of the most important events in the history of the twentieth-century life sciences, and discussions of how best to characterize the history of the synthesis period have remained fractious.[3] Stephen Jay Gould, paleontologist (turned philosopher, historian, and noted columnist), contended that the evolutionary synthesis was best described as a series of historical phases. The first phase, in the 1920s and 1930s, was the demonstration that Mendelian genetics was *theoretically* consistent with Darwinian evolutionary theory.[4] The second phase began with Dobzhansky's 1937 *Genetics and the Origin of Species*, as population geneticists and naturalists *experimentally* and *observationally* demonstrated the commensurability of genetics and evolution in wild organisms.[5] However, as Gould noted, even the primary actors in these two phases did not identify the "evolutionary synthesis" as a historical entity until the 1959 Darwin Centennial Celebration in Chicago. This conference solidified the identity of select evolutionary biologists as "synthesis architects."[6]

At the Chicago conference, evolutionary biologists and paleontologists recognized their agreement on a variety of key issues in evolutionary biology, from the proper level of biological organization at which to study evolution (the individual organism) to the importance of studying evolution in action. They further identified Darwin as their ancestral hero, thereby imbuing their community with some of Darwin's authority.[7] For Gould, the significance of the evolutionary synthesis rested not in its power to unify biologists' evolutionary theories before the Second World War, but in this later (third) phase of hardening or constricting the acceptable ways to conceive of evolution, culminating in the 1960s.[8] Gould devoted much of his career to breaking what he argued was an almost hegemonic control of the synthesis architects over evolutionary theory. He joined other paleobiologists and a new generation of evolutionary theorists in the 1970s who cast themselves as renegades, definitively breaking with the classical "neo-Darwinians" (Mayr and Dobzhansky, in particular) who had wrought the earlier hardening of evolutionary theory into a rigid adaptationist framework.[9] This rebellion could count as a fourth phase of the evolutionary synthesis. Historians' conceptions of the synthesis epoch are still being revised, but Gould's multiphasic history is helpful in characterizing the effects of evolutionary theory on the study of animal behavior throughout the twentieth century.

What, then, does the literature on the evolutionary synthesis have to say about the role of animal behavior research in our understanding of evolutionary theory (and vice versa)? Surprisingly, Ernst Mayr was one of the few people interested in the history of the synthesis to consider the place of animal behavior within the synthesis narrative.[10] According to Mayr, not all biologists embraced the centrality of the core synthesis texts in their understanding of biological processes. Besides those benighted molecular biologists, he claimed, biologists interested in the behavior or development of organisms were also particularly likely to ignore the conclusions of the evolutionary synthesis in their research. He suggested that both behavioral and developmental biologists emphasized change within an individual's lifetime, rather than evolutionary change of a population, and this had blinded them to the conclusions of the synthesists. At one of Mayr's 1974 conferences to explore the history of the evolutionary synthesis, young evolutionary biologist Robert Trivers suggested that behavior was integrated into the synthesis framework only with the work of William Hamilton and George Williams on kin selection in the mid-1960s and with Trivers's own theories of sexual selection in the early 1970s.[11]

Yet the differentiated histories of female choice and sexual selection illustrate how evolutionary theory and animal behavior were continuously resynthesized throughout the twentieth century. The first phase of the synthesis, the mathematical reconciliation of genetics and evolutionary theory, fitted genetic theories into existing ideas about how evolution worked in natural populations. Much like Darwin, evolutionary theorists in the early twentieth century believed that although natural selection might account for offspring *quantity*, offspring *quality* derived from the aesthetic process of mate choice. They argued that selection acting on both the quantity and quality of offspring through the processes of natural selection and sexual selection, respectively, would cause a single population to either improve or degenerate over time. A true sea change in the relationship between behavior and evolution came with the second phase of the synthesis in the late 1930s and 1940s. Biologists interested in behavior abandoned a progressive or degenerative model of evolution, with humans occupying the pantheon of rational behavior, and instead began to use a multidirectional model of evolution that emphasized the process of speciation. Within this new model of behavioral research, the mating behavior of any one individual was less important than the average mating behavior of the population.

Historians typically attribute this transition, from thinking about evolution in terms of behavioral and racial hierarchies to a branching, nonracially inflected model, to the mathematization of natural selection.[12] By redefining evolution in mathematical terms, theoretical population geneticists such as Ronald Fisher, J. B. S. Haldane, and Sewall Wright removed the taint of eugenics from natural selection by redefining fitness in nonphysical terms and transforming the agent of evolutionary change from differential survival to differential reproduction. Sociologist Howard Kaye, for example, has suggested that the mathematical phase of the modern synthesis softened "the harshness and brutality of the struggle for existence . . . [with] the aesthetic and sexual 'tastes' of the selecting sex," thereby avoiding "the undesirable implications of natural selection for human existence."[13] The evolutionary victor was the most attractive, not necessarily the strongest.

Yet it is only in retrospect that we can see the mathematics as divorced from the theory's eugenic implications. Fisher mathematized evolutionary theory so that it was consistent with his eugenic hopes and fears for human evolution. Although his conception of fitness was not based on the ability of individuals merely to survive, it remained a physical fitness nonetheless, in which the evolutionary success of individuals was measured by both their attractiveness to members of the opposite sex and the number of offspring they produced. By demonstrating that Mendelian inheritance was consistent with gradualist Darwinian evolution, Fisher saw the issues of heredity and evolution as mutually informative, and thus laid the foundations for his later mathematical formulations of reproduction as key to evolutionary progress.[14] He mathematically codified an aspect of his social agenda that differed from the sterilization campaigns associated with negative eugenics—an agenda based primarily on teaching people how to make the right kind of mate choices and encouraging eugenic couples to have more children.[15]

Fisher's history suggests that Darwinian and hereditarian arguments first melded in eugenic theory. Certainly, mechanisms such as sexual selection and social selection in humans played a far greater role in Fisher's theories of evolution than did the biological investigations of *Drosophila* researchers.[16] Although in his later writings Fisher did appeal to the experimental genetics literature and cite the work of fruit fly geneticists as important to the fusion of biometry and Mendelism, he barely mentioned *Drosophila* research in his 1930 *Genetical Theory of Natural Selection*.[17] In this book, Fisher discussed *Drosophila* research in only two contexts: the evolution of dominance and the dynamics of sex determination. When he did raise the topic of *Drosophila* research in the context of human evolution, he divided genetic researchers into two equally contemptible camps: those who thought their research said nothing about humans, and those who wished their research to say everything about human inheritance. Fisher argued that members of this second group were more prepared "to appreciate the similarities of human inheritance to that of *Drosophila* or Maize, than [they were] to appreciate the special problems which the evolution of man in society presents."[18] Sociological and historical facts were far more important than the work of laboratory geneticists in Fisher's formulation of evolutionary change. His intended implications of runaway sexual selection for human mate choice in *Genetical Theory* become clear: sexual selection was part of his larger commitment to fitness as a physical phenomenon determining the future of evolutionary change in human populations.[19]

The implications of evolutionary theory for the future of humanity became less straightforward in the late 1930s and 1940s, as biologists became more concerned with the process of speciation than with directional evolution in a single population. Gould's second phase of the synthesis, the appropriation of mathematical population genetics by naturalists, provided biologists with new perspectives on the process of evolution. Experimental population geneticists and systematists began to consider how speciation, the splitting of one interbreeding population into two reproductively isolated populations, might occur in nature. Using contemporary genetic techniques, this research program explained the origins of species diversity. Using Darwinian evolutionary theory, it also explained the maintenance of genetic variation in natural populations. The study of animal behavior as an evolutionary mechanism was radically transformed as a result of this new research program.

Curators of the Laboratory of Experimental Biology, later the Department of Animal Behavior, at the American Museum of Natural History (AMNH) sought to maintain funding for their facilities by directing their research at questions on the border of animal behavior and evolutionary theory. Thus, as biologists shifted from a progressive linear model of evolutionary change to a multidirectional, branching model, so did these curators shift their research from seeking to understand the evolutionary pattern of social and sexual behavior in animals to uncovering the relative importance of behavior as an evolutionary mechanism of speciation. In the 1930s, the founding curator of the laboratory, Gladwyn Kingsley Noble, had investigated the mating behavior of a variety of species, in the hope of constructing an evolutionary map of sexual behavior in animals, from fish and reptiles to mammals. His research on the hormonal and ecological causes of sexual behavior in these animals, he argued, would help him understand the social and sexual behavior of humans. Most of his experiments involved analyzing the mating behavior of two individuals at a time, as characteristic of the species to which they belonged. Later curators at the AMNH also analyzed the mating behavior of animals, but did so on a much grander scale. By the 1950s, even though they used some of the same organisms as Noble, Lester Aronson and his collaborators observed several hundred mating events in guppies and collated their data to look for statistical patterns in mate recognition across species boundaries. Re-creating the pattern of animal antecedents to human behavior was no longer part of the evolutionary biologists' research agenda; they sought, instead, to understand the relative importance of female choice as a mechanism to maintain reproductive isolation between different species.

In his early research at the AMNH, Aronson set about incorporating a population-level approach into the study of animal behavior that stood in stark contrast to most contemporary research in comparative psychology. Mayr, one of Aronson's fellow curators at the AMNH, disparaged most comparative psychology as involving the same rat in two different mazes.[20] Mayr commended the recent work of Konrad Lorenz and Niko Tinbergen, but insisted that what behavioral biologists needed to do in the future was analyze the behavior of entire populations. Population-level analyses would allow behavioral biologists to understand behaviors as taxonomic characters and the process of how natural selection acted to modify behavior. Aronson hoped that his research on guppies would fit well within this framework.

In the 1950s and 1960s, two communities of scientists embraced such population-based approaches to animal behavior in rather distinct research programs—ethologically trained zoologists and population geneticists. Although both groups saw their research as elucidating the modern plight of humanity, each approach to behavior implied a different kind of model system. On the one hand, ethologist Niko Tinbergen suggested that the same methodology ethologists applied to animals could be applied to understanding human behavior. To understand the cause, function, ontogeny, and evolution of human behaviors, zoologists needed to carefully observe people behaving in their natural environment. Other ethologists used this approach to reveal the ecological and evolutionary advantages of universal human behaviors. Humans, then, represented not a pinnacle of behavioral evolution, but one species' unique set of evolutionary adaptations to its rather unusual environment. On the other hand, population geneticists used *Drosophila* as an easily manipulable genetic model for humans. Fruit flies, like humans, were cosmopolitan species that sustained high levels of genetic variability over wide geographic ranges. Whereas ethologists tended to use animals as model systems for experimental design, population geneticists used animals as stand-ins for humans.

The implications of ethology and population genetics for females' mate choice behavior differed considerably: ethologically trained zoologists advanced a threshold model of behavior, and population geneticists offered an active-choice model. The two models of female choice involved different experimental approaches to the systematic study of evolution and behavior, and presupposed different kinds of questions, subjects of analysis, levels of biological organization, and methodologies. Whereas ethologists asked how evolution acted to modify behavior, population geneticists were more concerned with how behavior could modify the evolutionary process. At its heart, this difference stemmed from the fact that ethologists saw behavior as the primary focus of their research, while population geneticists sought to understand the process of evolution.

In the 1970s, while the disciplinary identity of female choice and sexual selection was still malleable, the study of animal behavior was once again resynthesized with evolutionary theory. Like Gould, the new generation of animal behavior researchers fought against the classical neo-Darwinism defended by Mayr, Dobzhansky, and Huxley. The result was a new hierarchy of behavioral complexity, once again with human social and sexual behavior as the ultimate goal of research on the behavior of animals. This most recent of the successive syntheses of animal behavior with evolutionary theory dramatically influenced our understanding of the history of female choice and sexual selection. Although animal behavior researchers integrated what they identified as the agenda of the evolutionary synthesists in the 1940s and 1950s, they did so in ways that biologists in the 1970s found difficult to recognize. This new generation of evolutionary biologists either couched earlier efforts to synthesize evolution and behavior as misguided, in the case of the neo-Darwinians, or conveniently ignored the politically distasteful contexts of Fisher's eugenic synthesis and Noble's hierarchical view of behavior.

Unpacking the changing dynamics of evolutionary theory and animal behavior allows us to see more clearly the disciplinary stakes and tools of legitimation inherent to historical narratives of animal behavior or evolutionary biology.

The history of female choice in evolutionary biology is a history we thought we knew—during the twentieth century, very few scientists framed their research in terms of Darwinian sexual selection, and when they talked about courtship behavior, their attention was focused on male-male competition. This narrative, however, stabilized during a time of disciplinary upheaval within the biological sciences and was framed by the dichotomies of great concern to biologists in the 1970s and 1980s: molecular versus organismal approaches, laboratory versus field research sites, and proximate versus ultimate causation. By broadening our historical vision to include theories of mate choice in humans, psychobiology, population genetics, and ethology, the history of female choice looks rather different. Laboratory and field research provided complementary approaches to understanding both the evolution of sexual behavior and the ways in which behavior could alter the process of evolution itself.

The history of biologists' research on female choice also elucidates the changing symmetries and divisions between theories of animal and human behavior. It seems obvious that there is much we can learn about ourselves from observing the natural world, and yet, throughout these years, no method for using behavioral or evolutionary knowledge of animals has gone unchallenged as a way of understanding human behavior. It is equally apparent that humans, with our rich cultural endowment, linguistic flexibility, technological know-how, and our sense of time, are unlike all other animals. One can make the argument, of course, that all living species differ in some distinct way from other organisms, but humans are perhaps uniquely concerned with what makes us different (or the same). Seeking to understand the sexual behavior of humans thus raises fundamental questions about who we are, from definitions of masculinity or femininity to the nature of choice.

Over the course of researching and writing, I have accrued a great many debts to many great people. This book began as a doctoral dissertation in the Department of the History of Science at the University of Wisconsin, and it is a great pleasure to thank my advisors, Gregg Mitman and Lynn Nyhart, for their encouragement, advice, and friendship. The history of science community in Madison made graduate school a true delight.

I also completed an M.S. in biology at the University of Michigan. My advisor there, Jerry Smith, encouraged me to think broadly about paleontology, animal behavior, and evolutionary theory, including their historical development. John Carson took a chance on a biologist and introduced me to the wonders of the history of science.

While researching this project, I visited many archives on three different continents. The following archivists, curators, and collections managers with whom I worked were tremendously helpful: John Henry Bennett and Cheryl Hoskin (Barr Smith Library of the University of Adelaide), M. Linda Birch (Zoology and Alexander Librarian, Oxford University Library Services), Mandy York Focke (Woodson Research Center, Rice University), Michael Gabriel (Roosevelt University), Janice Goldblum (National Academies Archives), Colin Harris (Department of Special Collections and Western Manuscripts, Bodleian Library, University of Oxford), Barbara Mathé (Special Collections, American Museum of Natural History [AMNH] Research Library), Barbara Meloni (Pusey Library, Harvard University Archives), Chuck Myers, Darrell Frost, and David Dickey (Department of Herpetology, AMNH), Tracy Elizabeth Robinson (Smithsonian Institution Archives), and Ethel Tobach (Department of Mammalogy, AMNH). I also spent a highly productive and enjoyable month and a half at the American Philosophical Society (APS) Library, thanks to Joseph James Ahern, Valerie-Ann Lutz, Roy Goodman, and Charles Greifenstein.

In addition, Lee Ehrman, Claudine Petit, John Maynard Smith, Eliot Spiess, and Robert Trivers graciously shared memories of their scientific research and colleagues, and I thank them for their time and thoughtfulness. Nicolas Gompel put me in touch with Mme. Claudine Petit.

Colleagues at Clemson University were invaluable as resources and friends as I began to think about turning the dissertation into a book. Thanks to Jake Hamblin, Pam Mack, Tom Oberdan, and the late Jerry Waldvogel for nurturing the Science and Technology in

Society community on campus. Thanks also to the members of the Biology Department who listened to my ideas and generously offered their thoughts and encouragements, especially Carroll Belser, Rick Blob, Bryan Brown, Michael Childress, Saara DeWalt, Nora Espinoza, Sid Gauthreaux, John Hains, Karen Hall, Kalan Ickes, Kim Paul, Margaret Ptacek, and Lisa Rapaport.

Writing, I have learned, takes time, and without the opportunity to spend a year and a half at the Max Planck Institute for the History of Science in Berlin, this book would have been longer in coming and almost certainly less interesting. I thank Lorraine Daston for taking me on as a postdoctoral fellow, and the people who spent time in and around Department II for creating a welcoming and intellectually stimulating environment in which to work. I owe special thanks to Charlotte Bigg, Luis Campos, Michael Gordin, Philip Kitcher, Maria Kronfeldner, Nick Langlitz, Daryn Lehoux, Andreas Mayer, Elysse Newman, Christine von Oertzen, Christian Reiß, Skúli Sigurdsson, Thomas Sturm, Kelly Wilder, Annette Vogt, and, of course, Tania Munz, for many delightful conversations about animals, gender, and science.

As I wrote, and rewrote, I shared drafts of my work with many kind souls, whose acute suggestions clarified and improved my arguments and prose beyond measure. I thank the following people for reading those drafts (some of which were quite rough), either in part or whole: John Beatty, Jenny Boughman, Joe Cain, Lorraine Daston, Paul Erickson, Libbie Freed, Fred Gibbs, Michael Gordin, Jake Hamblin, Judy Houck, Maria Kronfeldner, Joshua Kundert, Pam Mack, Gregg Mitman, Lynn Nyhart, Brent Ruswick, Jonathan Seitz, Elliott Sober, Fernando Vidal, Stephen Wald, Brett Walker, Kelly Whitmer, and Kärin Zachmann. As I neared a final draft, Chip Burkhardt and Tania Munz read through the entire manuscript long after I was capable of seeing anything with fresh eyes. At the crucial moments when this project congealed into text, Brad Hersh shaped both its content and its form in too many ways to count.

I feel lucky to be part of the History Department of the University of Maryland. Thanks especially to the Maryland Colloquium for the History of Science, Technology, and the Environment, and to the Washington, DC, History and Philosophy of Biology group for many convivial discussions, and the promise of so many more.

I would also like to acknowledge generous financial support for my research from several sources. At Wisconsin, I received a John Neu Wisconsin Distinguished Graduate Fellowship and a University of Wisconsin Dissertation Fellowship that provided me with, respectively, time to work in the early days of research and time to write during the final stages of my dissertation. A Library Resident Research Fellowship from the Friends of the American Philosophical Society Library let me explore at leisure their resources in the history of genetics and evolution. A Doctoral Dissertation Improvement Grant from the National Science Foundation (SES-0423612) enabled both my visits to other archives and my oral history interviews.

A modified version of chapter 3 appeared in Joe Cain and Michael Ruse's edited collection *Descended from Darwin: Insights into the History of American Evolutionary Studies, 1900–1970* (Philadelphia: Transactions of the American Philosophical Society, 2009), and I thank them for permission to include that material here.

I gratefully acknowledge permission to quote from the letters of Ronald Fisher (Barr

Smith Library at the University of Adelaide), the Julian Huxley Papers (Fondren Library, Woodson Research Center, Rice University), the Gladwyn Kingsley Noble and Charles Bogert Papers (Department of Herpetology, AMNH), the Ernst Mayr Papers (Harvard University Archives), the Department of Animal Behavior records (Special Collections, AMNH Research Library), the Dobzhansky Papers (APS), the Transcripts of the Evolutionary Synthesis Conference (APS), the Alan John Marshall Papers (National Library of Australia), and the National Research Council's Committee for Research in Problems of Sex (NRC-CRPS) papers (National Academies Archives). John Alcock granted me permission to quote from a letter of his to Ernst Mayr. The Tinbergen family kindly gave me permission to quote from the Nikolaas Tinbergen Papers.

Thanks also to Robert J. Brugger and Josh Tong for smoothly shepherding my manuscript through the review and publication process at the Johns Hopkins University Press. Harriet Ritvo and an anonymous reviewer offered numerous helpful suggestions for revising the manuscript. Linda Strange's keen editorial eye saved me from not a few convoluted misunderstandings.

My friends and family have sustained me throughout this process, listened patiently, plied me with wine and the occasional chocolate bar, laughed at my terrible jokes, and in general kept me sane. Shannon Withycombe, Fred Gibbs, Hilary Copp, Robert von Thaden, Fred Wuensch, Erin Engle, Faith Handley, and William Milam—I love you and thank you all from the bottom of my heart.

The following abbreviations are used for archival sources.

AMNH–Special Collections	Special Collections, American Museum of Natural History Research Library, New York
AMNH-Herpetology	Departmental Archives, Department of Herpetology, American Museum of Natural History, New York
APS	Manuscript Collections, American Philosophical Society, Philadelphia
AU-BSL	Special Collections, Barr Smith Library, University of Adelaide, Adelaide, Australia
HU	Harvard University Archives, Cambridge, Massachusetts
NLA	National Library of Australia, Canberra
NRC-DIV-MED-CRPS	NRC-CRPS Papers, 1946–52, National Research Council, Division of Medical Sciences, Committee for Research in Problems of Sex, 1920–1965, National Academies Archives, Washington, DC
OU-Bodleian	Bodleian Library, Department of Western Manuscripts, University of Oxford, Oxford
RU-Fondren	Woodson Research Center, Fondren Library, Rice University, Houston, Texas

Introduction

1. Darwin's *Origin of Species* contained a brief introduction to sexual selection, describing both competition between males and selection by females. Yet the only occurrence of the word *choice* in *Origin of Species* is in the first chapter, where Darwin discusses artificial selection and the breeding of "choice animals." He did not describe "choice exerted by the female" as an evolutionary mechanism until 1871, and he never used the phrase "female choice." We owe that, I believe, to Alfred Russel Wallace. Charles Robert Darwin, *On the Origin of Species by Means of Natural Selection or the Preservation of Favoured Races in the Struggle for Life*

(London: John Murray, 1859); Charles Robert Darwin, *Descent of Man and Selection in Relation to Sex* (London: John Murray, 1871), 2 vols.

2. Mary Bartley, "Conflicts in Human Progress: Sexual Selection and the Fisherian 'Runaway,'" *British Journal for the History of Science* 27 (1994); Mary Bartley, "Courtship and Continued Progress: Julian Huxley's Studies on Bird Behavior," *Journal of the History of Biology* 28 (1995); Helena Cronin, *The Ant and the Peacock: Altruism and Sexual Selection from Darwin to Today* (Cambridge: Cambridge University Press, 1991); Simon J. Frankel, "The Eclipse of Sexual Selection Theory," in *Sexual Knowledge, Sexual Science: The History of Attitudes to Sexuality*, ed. Roy Porter and M. Teich (Cambridge: Cambridge University Press, 1994). Cronin's readable and engaging *Ant and the Peacock* started me thinking about the history of sexual selection theory. In it, she investigated the philosophical stakes in the history of altruism (ant) and sexual selection (peacock). In terms of sexual selection, she paid particular attention to the resonance between the kinds of research questions posed by biologists in recent years and the questions posed by the debates over female choice in the time of Darwin and Wallace, a century earlier. The major difference between the two, she posited, was anthropomorphism. Darwin and Wallace read their cultural convictions into the emotions and choices of their subjects, whereas modern biologists had stripped away anthropomorphic language from their research and stood on much stronger analytical ground. Although Cronin did not explicitly argue for an "eclipse" of sexual selection on historical grounds, her coverage of topics certainly implied that one had taken place. Both Bartley and Frankel expanded Cronin's analysis in interesting ways. Bartley's work highlights the importance of Julian Huxley's and Ronald Fisher's assumptions about human evolution and sexuality in their understanding of sexual selection as a theory. In Frankel's article based on his unpublished M.Phil. thesis, he echoes Cronin in arguing that anthropomorphic notions of choice made sexual selection unpalatable to most biologists, but he also draws attention to the role of population geneticists in providing non-anthropomorphic venues for sexual selection research in the late 1960s.

3. Robert L. Trivers, "Parental Investment and Sexual Selection," in *Sexual Selection and the Descent of Man, 1871–1971*, ed. Bernard Clark Campbell (Chicago: Aldine Publishing Co., 1972).

4. Philip J. Pauly, *Controlling Life: Jacques Loeb and the Engineering Ideal in Biology*, Monographs in History and Philosophy of Science (Oxford: Oxford University Press, 1987); Daniel P. Todes, *Pavlov's Physiology Factory: Experiment, Interpretation, Laboratory Enterprise* (Baltimore: Johns Hopkins University Press, 2001).

5. Bert Bender, *Evolution and "The Sex Problem": American Narratives during the Eclipse of Darwinism* (Kent, OH: Kent State University Press, 2004); Bert Bender, *The Descent of Love: Darwin and the Theory of Sexual Selection in American Fiction, 1871–1926* (Philadelphia: University of Pennsylvania Press, 1996); Wendy Kline, *Building a Better Race: Gender, Sexuality, and Eugenics from the Turn of the Century to the Baby Boom* (Berkeley: University of California Press, 2001); Angelique Richardson, *Love and Eugenics in the Late Nineteenth Century: Rational Reproduction and the New Woman* (New York: Oxford University Press, 2003).

6. William B. Provine, *Origins of Theoretical Population Genetics* (Chicago: University of Chicago Press, 1971; reprint, 2001); William B. Provine, *Sewall Wright and Evolutionary Biology* (Chicago: University of Chicago Press, 1986).

7. See, for example, Ronald Aylmer Fisher, *The Genetical Theory of Natural Selection*

(Oxford: Clarendon Press, 1930); J. B. S. Haldane, *The Causes of Evolution* (Princeton, NJ: Princeton University Press, 1932; reprint, 1990, with new introduction and afterword by Egbert G. Leigh Jr.); Sewall Wright, "Evolution in Mendelian Populations," *Genetics* 16 (1931).

8. Susan M. Rensing, "Feminist Eugenics in America: From Free Love to Birth Control, 1880–1930" (Ph.D. diss., University of Minnesota, 2006).

9. Joseph Allen Cain, "Common Problems and Cooperative Solutions: Organizational Activity in Evolutionary Studies," *Isis* 84 (1993); Joseph Allen Cain, "Ernst Mayr as Community Architect: Launching the Society for the Study of Evolution and the Journal *Evolution*," *Biology and Philosophy* 9 (1994); Joseph Allen Cain, "Towards a 'Greater Degree of Integration': The Society for the Study of Speciation, 1939–41," *British Journal for the History of Science* 33 (2000); Ernst Mayr and William B. Provine, eds., *The Evolutionary Synthesis: Perspectives on the Unification of Biology* (Cambridge, MA: Harvard University Press, 1980); Vassiliki Betty Smocovitis, "Disciplining Evolutionary Biology: Ernst Mayr and the Founding of the Society for the Study of Evolution and *Evolution* (1939–1950)," *Evolution* 48, no. 1 (1994); Vassiliki Betty Smocovitis, "Keeping Up with Dobzhansky: G. L. Stebbins, Plant Evolution and the Evolutionary Synthesis," *History and Philosophy of the Life Sciences* 28 (2006); Vassiliki Betty Smocovitis, *Unifying Biology: The Evolutionary Synthesis and Evolutionary Biology* (Princeton, NJ: Princeton University Press, 1996).

10. Richard W. Burkhardt Jr., *Patterns of Behavior: Konrad Lorenz, Niko Tinbergen, and the Founding of Ethology* (Chicago: University of Chicago Press, 2005). Although Karl von Frisch's research was not truly ethological in nature, his investigations into the honey bee's waggle dance seemed to connect animal and human minds through a semantic language. Tania Munz, "Of Birds and Bees: Karl von Frisch, Konrad Lorenz and the Science of Animals, 1908–1973" (Ph.D. diss., Princeton University, 2007); Tania Munz, "The Bee Battles: Karl von Frisch, Adrian Wenner and the Honey Bee Dance Language Controversy," *Journal of the History of Biology* 38 (2005).

11. Although rare in usage outside art history, the term *zoomorphism* refers to the process of attributing animal forms to gods or superhuman beings. It was also used, as I intend here, by Ethel Tobach in 1976 to describe the sociobiological "reduction of humans to lower levels of organization." Ethel Tobach, "Evolution of Behavior and the Comparative Method," *International Journal of Psychology* 11 (1976): 195.

12. Debates over primates' capacity for language invoked similar concerns among biologists about anthropomorphism and professional identity. For a fascinating account of this history, see Gregory Radick, *Simian Tongue: The Long Debate about Animal Language* (Chicago: University of Chicago Press, 2007).

Chapter 1 · *Beauty and the Beast*

Epigraph. Richard Grant White, *The Fall of Man: Or, the Loves of the Gorillas. A Popular Scientific Lecture upon the Darwinian Theory of Development by Sexual Selection, by a Learned Gorilla* (New York: G. W. Carleton & Co., Publishers, 1871), 8.

1. John Gould, *Handbook to the Birds of Australia*, vol. 1 (London: Published by the Author, 1865), 444. Darwin cited this volume extensively in *Descent of Man and Selection in Relation to Sex* (London: John Murray, 1871), 2 vols.

2. Janet Browne, *Charles Darwin: The Power of Place, Volume II of a Biography* (New York: Alfred A. Knopf, 2002).

3. John Durant, "The Ascent of Nature in Darwin's *Descent of Man*," in *The Darwinian Heritage,* ed. David Kohn (Princeton, NJ: Princeton University Press, 1985); Malcolm Jay Kottler, "Charles Darwin and Alfred Russel Wallace: Two Decades of Debate over Natural Selection," in Kohn, *Darwinian Heritage*; Malcolm Jay Kottler, "Darwin, Wallace, and the Origin of Sexual Dimorphism," *Proceedings of the American Philosophical Society* 124, no. 3 (1980).

4. Erasmus Darwin, *The Temple of Nature; or, the Origin of Society: A Poem with Philosophical Notes* (London: Jones and Co., 1825), 13, 52.

5. For more on the life of Charles Darwin, see Browne, *Charles Darwin: Power of Place*; Janet Browne, *Charles Darwin: Voyaging, Volume I of a Biography* (New York: Alfred A. Knopf, 1995); Adrian Desmond and James Moore, *Darwin* (London: Joseph, 1991).

6. Charles Robert Darwin, *On the Origin of Species by Means of Natural Selection or the Preservation of Favoured Races in the Struggle for Life* (London: John Murray, 1859), 87–89. Apart from the origins of secondary sexual characteristics, Darwin's discussion of sexual selection (156–58) recurs with regard to rudimentary organs (197–99).

7. Ibid., 88, 89.

8. Darwin, *Descent of Man* (1871), vol. 1, 240–50. In these pages, Darwin argued that "the characteristic differences between the races of man ["colour, hairyness, form of features" (250)] cannot be accounted for in a satisfactory manner by the direct action of the conditions of life [natural selection], nor by the effects of the continued use of parts, nor through the principle of correlation." However, he continued, "there remains one important agency, namely Sexual Selection, which appears to have acted as powerfully on man, as on many other animals" (248, 249). This passage forms the bridge between Part I, "On the Descent of Man," and Part II, "Sexual Selection." Adrian Desmond and James Moore, *Darwin's Sacred Cause: How a Hatred of Slavery Shaped Darwin's Views on Human Evolution* (New York: Houghton Mifflin, 2009).

9. George John Romanes, *Darwin and after Darwin: An Exposition of the Darwinian Theory and a Discussion of Post-Darwinian Questions,* vol. I: *The Darwinian Theory* (Chicago: Open Court Publishing Co., 1892), 277, and further elaborated in chap. X, "The Theory of Sexual Selection and Concluding Remarks."

10. Durant, "Ascent of Nature in Darwin's *Descent of Man*," 297.

11. Charles Robert Darwin, *The Descent of Man, and Selection in Relation to Sex,* 2nd ed. (London: John Murray, 1882), 211. A slightly wordier version of this sentence can be found in *Descent of Man* (1871), vol. 1, 259.

12. Darwin, *Descent of Man* (2nd ed., 1882), 211. Darwin's comparison of the "taste for the beautiful in animals" with that of the "lowest savages" was a point of clarification he added in this second edition.

13. Darwin, *Descent of Man* (1871), vol. 1, 273.

14. Darwin, *Descent of Man* (2nd ed., 1882), 617. See also Richard W. Burkhardt Jr., "Darwin on Animal Behavior and Evolution," in Kohn, *Darwinian Heritage.*

15. Darwin, *Descent of Man* (2nd ed., 1882), 589.

16. Ibid., 220 (fishes), 319 (butterflies), 406, 495 (birds), 522 (quadrupeds), 573–85 (mankind).

17. Darwin, *Descent of Man* (1871), vol. 2, 122.

18. Evelleen Richards, "Darwin and the Descent of Woman," in *Wider Domain of Evolutionary Thought*, ed. David Oldroyd and Ian Langham (London: D. Reidel Publishing Co., 1983).

19. Darwin, *Descent of Man* (1871), vol. 2, 316, 326, 326–27, 328–29.

20. Browne, *Charles Darwin: Power of Place*, 344–46; Burkhardt, "Darwin on Animal Behavior and Evolution," 349; James Moore and Adrian Desmond, "Introduction," in *Charles Darwin: The Descent of Man* (London: Penguin Books, 2004), xlviii.

21. Darwin, *Descent of Man* (1871), vol. 1, chap. VII, "On the Races of Man"; vol. 2, chap. XIX, "Secondary Sexual Characters of Man"; and chap. XX, "Secondary Sexual Characters of Man—*continued.*"

22. Ibid., vol. 1, 241–48.

23. Ibid., vol. 1, 241, 250; vol. 2, 383–84; Moore and Desmond, "Introduction," xii, xxv–xxx; Desmond and Moore, *Darwin's Sacred Cause.*

24. Darwin, *Descent of Man* (2nd ed., 1882), 617.

25. Ibid., 617.

26. Ibid., 573, 577–80; Cynthia Eagle Russett, *Sexual Science: The Victorian Construction of Womanhood* (Cambridge, MA: Harvard University Press, 1989), 80.

27. Darwin, *Descent of Man* (2nd ed., 1882), 617.

28. Browne, *Charles Darwin: Power of Place*, 368–69; Janet Browne, "Darwin and the Expression of the Emotions," in Kohn, *Darwinian Heritage*; Charles Robert Darwin, *The Expression of the Emotions in Man and Animals* (London: John Murray, 1872), 78–79.

29. Darwin, *Expression of the Emotions*, 215–16. Darwin stated that although many people would assume that kissing was a purely human trait, some chimpanzees were known to touch their lips together in displays of mutual affection, and kissing was unknown in a variety of human cultures.

30. Ibid., 84.

31. Ibid., 87; Darwin, *Descent of Man* (1871), vol. 2, 330–37.

32. Darwin, *Origin of Species* (1871); Darwin, *Expression of the Emotions*; St. George Mivart, "*The Descent of Man and Selection in Relation to Sex*. By Charles Darwin, M.A., F.R.S., &c. 2 vols. London, 1871" (review), *Quarterly Review* 131, July (1871); Robert J. Richards, *Darwin and the Emergence of Evolutionary Theories of Mind and Behavior* (Chicago: University of Chicago Press, 1987); Russett, *Sexual Science.*

33. Alfred Russel Wallace, "The Origin of Human Races and the Antiquity of Man Deduced from the Theory of 'Natural Selection,'" *Journal of the Anthropological Society of London* 2 (1864): clxvi.

34. Ibid., clxviii; A.D., "A Visit to Dr. Alfred Russel Wallace, F.R.S.," *Bookman (London)* 13 (1898).

35. Wallace, "Origin of Human Races," clxvi; John R. Durant, "Scientific Naturalism and Social Reform in the Thought of Alfred Russel Wallace," *British Journal for the History of Science* 12 (1979).

36. Wallace, "Origin of Human Races," clxv.

37. Thomas F. Glick, "The Anthropology of Race across the Darwinian Revolution," in *A New History of Anthropology*, ed. Henrika Kuklick (Oxford: Blackwell Publishing, 2008); George W. Stocking Jr., *Race, Culture, and Evolution: Essays in the History of Anthropology* (New York: Free Press, 1968), chap. 3, "The Persistence of Polygenist Thought in Post-Darwinian Anthropology."

38. Durant, "Scientific Naturalism," 40–42; Wallace, "Origin of Human Races," clxvi.

39. Alfred Russel Wallace, "The Limits of Natural Selection as Applied to Man," in *Contributions to the Theory of Natural Selection, a Series of Essays* (London: Macmillan and Co., 1870), 333, 339.

40. Durant, "Ascent of Nature in Darwin's *Descent of Man*," 290.

41. Wallace, "Limits of Natural Selection as Applied to Man," 341, 353.

42. Ibid., 350, 359.

43. A.D., "Visit to Dr. Alfred Russel Wallace."

44. Alfred Russel Wallace, "The Colors of Animals and Plants (part 1)," *American Naturalist* 11, no. 11 (1877); Alfred Russel Wallace, "The Colors of Animals and Plants (part 2)," *American Naturalist* 11, no. 12 (1877). Both originally published in *Macmillan's Magazine*, Sept. and Oct. 1877. See n. 4, p. 474, in *Macmillan's Magazine*, not present in *American Naturalist* version.

45. Wallace, *Macmillan's Magazine*, Sept. and Oct. 1877, 401–2.

46. Alfred Russel Wallace, *Darwinism*, 2nd ed. (London: Macmillan and Co., 1889), 273–81.

47. Alfred Russel Wallace, "Human Selection," *Fortnightly Review* 48 (1890).

48. Ibid., 325.

49. Ibid., 328, 331.

50. Alfred Russel Wallace, *Social Environment and Moral Progress* (New York: Cassell and Co., 1913), 151–53, emphasis in original.

51. Wallace, "Human Selection," 335–36, 337.

52. Wallace, *Social Environment*, 140–41.

53. Jean Gayon, *Darwinism's Struggle for Survival: Heredity and the Hypothesis of Natural Selection*. Cambridge Studies in Philosophy and Biology (Cambridge: Cambridge University Press, 1998), 514.

54. Karl Pearson, *The Life, Letters and Labours of Francis Galton*, vol. 3a: *Correlation, Personal Identification and Eugenics* (Cambridge: Cambridge University Press, 1930), chap. XIV, "Correlation and the Application of Statistics to the Problems of Heredity."

55. Karl Pearson, "Mathematical Contributions to the Theory of Evolution. III. Regression, Heredity, and Panmixia," *Philosophical Transactions of the Royal Society of London, Series A Mathematical* 187 (1896): 257–58.

56. George John Romanes, *Animal Intelligence* (London: Kegan Paul, Trench, 1881); George John Romanes, *Mental Evolution in Animals, with a Posthumous Essay on Instinct by Charles Darwin* (London: Kegan Paul, Trench, 1883).

57. Romanes, *Mental Evolution in Animals*, 20, emphasis in original.

58. Romanes, *Animal Intelligence*, 281.

59. Patrick Geddes and J. Arthur Thompson, *The Evolution of Sex* (London: Walter Scott, 1889; rev. ed. 1901).

60. Russett, *Sexual Science*, 89–92, 135–36.

61. Geddes and Thompson, *Evolution of Sex* (rev. ed., 1901), 297, in chap. XIX, "Psychological and Ethical Aspects."

62. Barbara T. Gates and Ann B. Shteir, eds., *Natural Eloquence: Women Reinscribe Science* (Madison: University of Wisconsin Press, 1997), 51.

63. Penelope Deutscher, "The Descent of Man and the Evolution of Woman," *Hypatia* 19, no. 2 (2004); Barbara T. Gates, *Kindred Nature: Victorian and Edwardian Women Embrace the Living World* (Chicago: University of Chicago Press, 1998); Sally Gregory Kohlstedt and Mark R. Jorgensen, "'The Irrepressible Woman Question': Women's Response to Darwinian Evolutionary Ideology," in *Disseminating Darwinism: The Role of Place, Race, Religion, and Gender*, ed. Ronald L. Numbers and John Stenhouse (Cambridge: Cambridge University Press, 1999).

64. Eliza Burt Gamble, *The Evolution of Woman: An Inquiry into the Dogma of Her Inferiority to Man* (New York: G. P. Putman's Sons, 1894). The second edition of this book (1916), also published as *The Sexes in Science and History*, was substantially unaltered.

65. Mike Hawkins, *Social Darwinism in European and American Thought, 1860–1945: Nature as Model and Nature as Threat* (New York: Cambridge University Press, 1997), 261.

66. Gamble, *Evolution of Woman* (2nd ed., 1916), 31, 75.

67. Ibid., 380.

68. Gilman features prominently in many histories of America feminism and women's health, only a few of which I cite here. Nancy F. Cott, "Passionlessness: An Interpretation of Victorian Sexual Ideology, 1790–1850," *Signs* 4, no. 2 (1978); Mary A. Hill, *The Making of a Radical Feminist, 1860–1896* (Philadelphia: Temple University Press, 1983); Ann J. Lane, *To Herland and Beyond: The Life and Work of Charlotte Perkins Gilman* (New York: Pantheon Books, 1997).

69. Charlotte Perkins Gilman, *Women and Economics: A Study of the Economic Relation between Men and Women as a Factor in Social Evolution* (Boston: Small, Maynard and Co., 1898).

70. Gates, *Kindred Nature*; Hawkins, *Social Darwinism in European and American Thought*; Hill, *Making of a Radical Feminist*, 268; Lane, *To Herland and Beyond*, 283; Angelique Richardson, *Love and Eugenics in the Late Nineteenth Century: Rational Reproduction and the New Woman* (New York: Oxford University Press, 2003), 76; Russett, *Sexual Science*.

71. Gilman, *Women and Economics*; Alice S. Rossi, ed., *The Feminist Papers: From Adams to de Beauvoir* (New York: Bantam Books, 1973). Gilman's view on the superiority of women is especially clear in her utopian novel *Herland*, originally published in serial form (1915). Charlotte Perkins Gilman, *Herland* (New York: Pantheon Books, 1979). In *Woman and Labour*, Olive Schreiner also criticized Victorian women as social parasites. Schreiner, *Woman and Labour*, Virago Reprint Library (London: Virago, 1978 [1911]).

72. George Robb, "Eugenics, Spirituality, and Sex Differentiation in Edwardian England: The Case of Frances Swiney," *Journal of Women's History* 10, no. 3 (1998).

73. Frances Swiney, *The Awakening of Women or Woman's Part in Evolution* (London: George Redway, 1899); Richardson, *Love and Eugenics*, 48.

74. Swiney, *Awakening of Women*, 20.

75. Robb, "Eugenics, Spirituality, and Sex Differentiation"; Swiney, *Awakening of Women*, 92–102.

76. Gates, *Kindred Nature*; Gates and Shteir, *Natural Eloquence*; Richardson, *Love and Eugenics*.

77. Hill, *Making of a Radical Feminist*, 268.

78. Patrick Parrinder, "Eugenics and Utopia: Sexual Selection from Galton to Morris," *Utopian Studies* 8, no. 2 (1997).

79. Sally Gregory Kohlstedt, "Nature Not Books: Scientists and the Origins of the Nature Study Movement in the 1890s," *Isis* 96, no. 3 (2005); Jennifer Mason, *Civilized Creatures: Urban Animals, Sentimental Culture, and American Literature, 1850–1900* (Baltimore: Johns Hopkins University Press, 2005).

80. Ralph Lutts, *The Nature Fakers: Wildlife, Science, and Sentiment* (Golden, CO: Fulcrum Publishing, 1990), chap. 5, "In Search of an Earthly Eden"; Jed Mayer, "The Expression of the Emotions in Man and Laboratory Animals," *Victorian Studies* 50, no. 3 (2008).

81. Cora Daisy Reeves, "The Breeding Habits of the Rainbow Darter (*Etheostoma coeruleum* Storer): A Study in Sexual Selection," *Biological Bulletin* 14 (1907); Cora Daisy Reeves, "Discrimination of Light of Different Wave-Lengths by Fish," *Behavior Monographs* 4, no. 3 (1919); Cora Daisy Reeves, "Moving and Still Lights as Stimuli in a Discrimination Experiment with White Rats," *Journal of Animal Behavior* 7, no. 3 (1917).

82. William C. Kimler, "Reading Morgan's Canon: Reduction and Unification in the Forging of a Science of the Mind," *American Zoologist* 40, no. 6 (2000); Gregory Radick, "Morgan's Canon, Garner's Phonograph, and the Evolutionary Origins of Language and Reason," *British Journal for the History of Science* 33 (2000).

83. Conway Lloyd Morgan, *Animal Life and Intelligence* (Boston: Ginn and Co., Publishers, 1891).

84. Ibid., 409, 460–61; Conway Lloyd Morgan, "Mental Factors in Evolution," in *Darwin and Modern Science: Essays in Commemoration of the Centenary of the Birth of Charles Darwin and of the Fiftieth Anniversary of the Publication of the Origin of Species*, ed. A. C. Seward (Cambridge: Cambridge University Press, 1909).

85. Morgan, *Animal Life and Intelligence*, 484–85.

86. Morgan's canon dramatically influenced psychologist Edward Thorndike, who in turn trained Robert Yerkes. Gregory Radick, *Simian Tongue: The Long Debate about Animal Language* (Chicago: University of Chicago Press, 2007), 8–9, 52, 201–20.

87. Robert J. Richards, *Darwin and the Emergence of Evolutionary Theories of Mind*. The *Journal of the History of Biology* published a special issue devoted to reevaluating historical conceptions of the importance (or lack thereof) of Darwinian mechanisms of evolutionary change at the turn of the century. "The 'Darwinian Revolution': Whether, What, and Whose?" *Journal of the History of Biology* 38, no. 1 (2005).

88. Robin Marantz Henig, *The Monk in the Garden: The Lost and Found Genius of Gregor Mendel, the Father of Genetics* (Boston: Houghton Mifflin Books, 2001); Philip J. Pauly, *Controlling Life: Jacques Loeb and the Engineering Ideal in Biology*, Monographs in History and Philosophy of Science (Oxford: Oxford University Press, 1987).

Chapter 2 · Progressive Desire

Epigraph. H. G. Wells, Julian S. Huxley, and G. P. Wells, *The Science of Life* (Garden City, NY: Doubleday, Doran and Co., 1931), vol. 2, 1239.

1. Some notable exceptions to this rejection of mate choice in animals were the early observations of Edmond Selous, William Pycraft, and Julian Huxley, whose research is addressed in this chapter, as well as Gladwyn Kingsley Noble at the American Museum of Natural History, whose research is discussed in chapter 3.

2. Cynthia Eagle Russett, *Sexual Science: The Victorian Construction of Womanhood* (Cambridge, MA: Harvard University Press, 1989).

3. For the importance of social hygiene and positive eugenics, see Daniel J. Kevles, *In the Name of Eugenics: Genetics and the Uses of Human Heredity* (New York: Alfred A. Knopf, 1985); Wendy Kline, *Building a Better Race: Gender, Sexuality, and Eugenics from the Turn of the Century to the Baby Boom* (Berkeley: University of California Press, 2001); Susan M. Rensing, "Feminist Eugenics in America: From Free Love to Birth Control, 1880–1930" (Ph.D. diss., University of Minnesota, 2006).

4. For the influence of human engineering as a concept in American social sciences, see Rebecca Lemov, *World as Laboratory: Experiments with Mice, Mazes, and Men* (New York: Hill and Wang, 2005).

5. Vernon Kellogg, *Headquarters Nights: A Record of Conversations and Experiences at the Headquarters of the German Army in France and Britain* (Boston: Atlantic Monthly Press, 1917), 29; Mark A. Largent, "Bionomics: Vernon Lyman Kellogg and the Defense of Darwin, 1890–1930," *Journal of the History of Biology* 32 (1999).

6. For a more nuanced view of German biological thought before the First World War, see Sander Gliboff, *H. G. Bronn, Ernst Haeckel, and the Origins of German Darwinism* (Cambridge, MA: MIT Press, 2008); Lynn K. Nyhart, *Biology Takes Form: Animal Morphology and the German Universities, 1800–1900* (Chicago: University of Chicago Press, 1995); Laura Otis, *Müller's Lab* (Oxford: Oxford University Press, 2007); Robert J. Richards, *The Tragic Sense of Life: Ernst Haeckel and the Struggle over Evolutionary Thought* (Chicago: University of Chicago Press, 2008).

7. The *New York Times* published twenty-six articles mentioning sexual selection between 1871 and 1900, and twenty-four between 1900 and 1939; in contrast, only two articles between 1940 and 1980 mentioned sexual selection. Of these, articles by or about George Bernard Shaw in the *New York Times* included the following: Marconi Transatlantic Wireless Telegraph to The New York Times, "English Can't Choose Mates, Says Shaw," Mar. 30, 1913; Shaw, "Shaw Expounds Socialism as World Panacea," Dec. 12, 1926; "Would Lift Marriage Bars" (Special to The New York Times), Feb. 11, 1928; Shaw, "The Case for Socialism," Sept. 12, 1926. The topic of sexual selection also arose numerous times in the *New York Times* in the context of the Scopes Trial: "Full Text of Mr. Bryan's Argument against Evidence of Scientists" (Special to The New York Times), July 17, 1925; "Text of Bryan's Evolution Speech, Written for the Scopes Trial," July 29, 1925; "Bryan Is Attacked before Scientists" (Special to The New York Times), Dec. 31, 1924; Henry Fairfield Osborn, "Evolution and Religion," Mar. 5, 1922; William Jennings Bryan, "God and Evolution," Feb. 26, 1922. Other *New York Times* articles

mentioning sexual selection included the following: Inis H. Weed, "What It Costs a Young Girl to Be Well Dressed," Apr. 24, 1910; "Man's Future in the Light of His Dim Past," Apr. 28, 1929; "Says Man Will Grow for Ages to Come" (Special to The New York Times), Apr. 20, 1929; Eugene S. Bagger, "Haldane Looks into the Future: A Review," Apr. 6, 1924.

8. "Would Force Rats into Race Suicide" (Special Correspondence of the New York Times), *New York Times*, Dec. 11, 1921.

9. With a few exceptions, research on animal courtship in the early decades of the twentieth century distinguished between humans and nonprimate animals. Primates as research organisms in their natural habitats were exceptionally hard to access until after the Second World War, and attempts to establish semi-natural primate-breeding colonies in locations closer to centers of research also proved difficult (in part because of the prohibitive expense). Kersten Jacobson Biehn, "Psychobiology, Sex Research and Chimpanzees: Philanthropic Foundation Support for the Behavioral Sciences at Yale University, 1923–41," *History of the Human Sciences* 21, no. 2 (2008); Donna Haraway, *Primate Visions: Gender, Race, and Nature in the World of Modern Science* (New York: Routledge, 1989); Georgina M. Montgomery, "Place, Practice and Primatology: Clarence Ray Carpenter, Primate Communication, and the Development of Field Methodology, 1931–1945," *Journal of the History of Biology* 38 (2005); Gregory Radick, *Simian Tongue: The Long Debate about Animal Language* (Chicago: University of Chicago Press, 2007).

10. Russett, *Sexual Science*.

11. Kellogg was an avid supporter of Darwinism in America. Vernon Kellogg, *Darwinism To-day: A Discussion of Present-day Scientific Criticism of the Darwinian Selection Theories, together with a Brief Account of the Principal Other Proposed Auxiliary and Alternative Theories of Species-Forming* (New York: Henry Holt and Co., 1907), 106–28, 48–50; Largent, "Bionomics."

12. Charles Robert Darwin, *The Descent of Man, and Selection in Relation to Sex*, 2nd ed. (London: John Murray, 1882), 243–60.

13. Kellogg, *Darwinism To-day*, 114.

14. J. R. Matthews, "History of the Cambridge Entomological Club," *Psyche* 81 (1974).

15. A. G. Mayer, "On the Mating Instinct in Moths," *Psyche* 9 (1900).

16. Wells, Huxley, and Wells, *Science of Life*, vol. 2, 1156. On animal reactions to the natural world, see Jakob von Uexküll, "A Stroll through the Worlds of Animals and Men: A Picture Book of Invisible Worlds" (first published in 1934), in *Instinctive Behavior*, ed. C. Schiller (New York: International Universities Press, 1957).

17. See, for example, David Starr Jordan and Vernon Lyman Kellogg, *Evolution and Animal Life: An Elementary Discussion of Fact, Processes, Laws and Theories Relating to the Life and Evolution of Animals* (New York: D. Appleton and Co., 1907); Kellogg, *Darwinism To-day*; Vernon Kellogg, *Evolution* (New York: D. Appleton and Co., 1924); Richard Swann Lull, *Organic Evolution* (New York: Macmillan Co., 1917); Wells, Huxley, and Wells, *Science of Life*.

18. Thomas Hunt Morgan, *Heredity and Sex* (New York: Columbia University Press, 1913); Thomas Hunt Morgan, *The Genetic and the Operative Evidence Relating to Secondary Sexual Characters*, Carnegie Institution of Washington pub. no. 285 (Washington, DC: Carnegie Institution of Washington, 1919); Thomas Hunt Morgan and H. D. Goodale, "Sex-Linked

Inheritance in Poultry," *Annals of the New York Academy of Sciences* 22 (1912). Thanks to Tania Munz for bringing Morgan's experiments on sexual selection in chickens to my attention.

19. Quoted in Kellogg, *Darwinism To-day*, 119. Amusingly, Ronald Aylmer Fisher was later to build up such a "fiction" (which he called "runaway sexual selection," discussed later in this chapter) astoundingly similar to Morgan's ironic account of female choice, and his theory is taken quite seriously today. A connection between the two is unlikely—certainly, Fisher does not cite Morgan. Malte Andersson and Yoh Iwasa, "Sexual Selection," *Trends in Ecology and Evolution* 11, no. 2 (1996); Mary Bartley, "Conflicts in Human Progress: Sexual Selection and the Fisherian 'Runaway,'" *British Journal for the History of Science* 27 (1994).

20. Sturtevant received his Ph.D. in 1914 and continued to work with Morgan at Columbia, funded by the Carnegie Institute of Washington, under whose jurisdiction this research was published in 1915. Edward B. Lewis, "Alfred Henry Sturtevant, November 21, 1891–April 5, 1970," in *Dictionary of Scientific Biography* (New York: Chas. Scribner's Sons, 1976). Also see Robert Kohler's wonderful book on the scientific and social dynamics of Morgan's *Drosophila* laboratory, including the relationship between Morgan (the "Boss") and the men, like Sturtevant, who worked in his laboratory (the "boys"). Robert Kohler, *Lords of the Fly: Drosophila Genetics and the Experimental Life* (Chicago: University of Chicago Press, 1994).

21. Alfred H. Sturtevant, "Experiments on Sex Recognition and the Problem of Sexual Selection in *Drosophila*," *Journal of Animal Behavior* 5 (1915): 366.

22. Today, such physiological explanations would not be considered incompatible with an evolutionary explanation; one could discuss, for example, both the specific biochemical pathways that make it possible for mammals to maintain a more or less constant body temperature in the bright sun of day and the cool dark of night *and* the selective pressures leading to such mechanisms of regulating body temperature in land mammals. For physiologists in the early twentieth century, however, a physiological explanation provided a simpler and less speculative cause of a trait than an evolutionary explanation. J. T. Cunningham, *Sexual Dimorphism in the Animal Kingdom: A Theory of the Evolution of Secondary Sexual Characters* (London: Black, 1900). Such physiological explanations were cited by both Vernon Kellogg and Thomas Hunt Morgan. Kellogg, *Darwinism To-day*, 118; Thomas Hunt Morgan, *The Scientific Basis of Evolution* (London: Faber and Faber, 1932), 159–60.

23. Alice Dreger, *Hermaphrodites and the Medical Invention of Sex* (Cambridge, MA: Harvard University Press, 1998), chap. 5, "The Age of Gonads."

24. Morgan, *Scientific Basis of Evolution*, chap. VII, "The Theory of Sexual Selection and Hormones."

25. Kellogg, *Darwinism To-day*, 149–50.

26. Ralph Lutts, *The Nature Fakers: Wildlife, Science and Sentiment* (Golden, CO: Fulcrum Publishing, 1990).

27. George W. Peckham and Elizabeth G. Peckham, "Additional Observations on Sexual Selection in Spiders of the Family Attidae, with Some Remarks on Mr. Wallace's Theory of Sexual Ornamentation," *Occasional Papers of the Natural History Society of Wisconsin* 1, no. 3 (1890); George W. Peckham and Elizabeth G. Peckham, "Observations on Sexual Selection in Spiders of the Family Attidae," *Occasional Papers of the Natural History Society of Wisconsin* 1, no. 1 (1889).

28. W. P. Pycraft, *The Courtship of Animals* (New York: Henry Holt and Co., 1913). In his

obituary, Pycraft was described as follows: "Being physically incapable of much active exercise, and never having very robust health, most of his work has naturally to be done indoors, otherwise he would have made a first-class field naturalist." G. E. Lodge, "Obituary: William Plane Pycraft," *Ibis* 85 (1943): 109.

29. Pycraft, *Courtship of Animals*, 146, xi–xiv.

30. Gladwyn Kingsley Noble to R. M. Strong, May 28, 1934, Noble Papers, folder: R. M. Strong, Departmental Archives, Department of Herpetology, American Museum of Natural History, New York (hereafter, AMNH-Herpetology).

31. Richard Burkhardt has written extensively on Selous's standing among professional zoologists in England, his passion for bird-watching, and his connection to future generations of British ornithologists through Julian Huxley. Richard W. Burkhardt Jr., "Edmund Selous," *Dictionary of Scientific Biography* 18, suppl. 2 (1990); Richard W. Burkhardt Jr., *Patterns of Behavior: Konrad Lorenz, Niko Tinbergen, and the Founding of Ethology* (Chicago: University of Chicago Press, 2005), 77–92.

32. Edmund Selous's publications included: "An Observational Diary of the Nuptial Habits of the Blackcock (*Tetrao tetrix*) in Scandinavia and England," *Zoologist* 13 (1909); "An Observational Diary of the Nuptial Habits of the Blackcock (*Tetrao tetrix*) in Scandinavia and England," *Zoologist* 14 (1910); *Bird Life Glimpses* (London: Allen, 1905); *Evolution of Habit in Birds* (London: Constable and Co., 1933); "Observations Tending to Throw Light on the Question of Sexual Selection in Birds, Including a Day to Day Diary on the Breeding Habits of the Ruff (*Machetes pugnax*)," *Zoologist* 10 (1906); *Realities of Bird Life: Being Extracts from the Diaries of a Life-Loving Naturalist* (London: Constable and Co., 1927); and "The Nuptial Habits of the Blackcock," *Naturalist* (1913).

33. Selous to R. A. Fisher, Nov. 9, 1932, Fisher Papers, Series I: Correspondence, MSS 0013, Special Collections, Barr Smith Library, University of Adelaide (hereafter, AU-BSL). Fisher had previously written to Selous congratulating him on the *Realities of Bird Life*, which Fisher regretted he had not read until after publication of his *Genetical Theory of Natural Selection*. Selous was complaining about what he saw as prejudicial attacks on his use of "sexual selection" in describing the birds' courtship behavior. Many, but not all, of Fisher's papers are now available online as part of the R. A. Fisher Digital Archive hosted by the University of Adelaide Digital Library, http://digital.library.adelaide.edu.au/coll/special/fisher/index.html.

34. Kenneth C. Waters and Albert Van Helden, eds., *Julian Huxley, Biologist and Statesman of Science: Proceedings of a Conference Held at Rice University, 25–27 September 1987* (Houston, TX: Rice University Press, 1992).

35. Mary Bartley, "Courtship and Continued Progress: Julian Huxley's Studies on Bird Behavior," *Journal of the History of Biology* 28 (1995); Burkhardt, *Patterns of Behavior*, 103–26.

36. Julian S. Huxley, *The Courtship Habits of the Great Crested Grebe*, ed. Nathaniel Tarn, Cape Editions, vol. 18 (London: Jonathan Cape, 1968 [1914]).

37. Ibid., 92.

38. Ibid.

39. Jordan and Kellogg, *Evolution and Animal Life*, 78.

40. Lull, *Organic Evolution*, 122, 101. Kellogg later repeated this point even more strongly. Kellogg, *Evolution*, 136.

41. Bartley, "Courtship and Continued Progress," 96–99.

42. Julian S. Huxley, "Darwin's Theory of Sexual Selection and the Data Subsumed by It, in the Light of Recent Research," *American Naturalist* 72 (1938); Julian S. Huxley, "The Present Standing of the Theory of Sexual Selection," in *Evolution: Essays on Aspects of Evolutionary Theory*, ed. Gavin de Beer (London: Oxford University Press, 1938).

43. Huxley, "Darwin's Theory of Sexual Selection," 418.

44. Julian S. Huxley, Ludwig Koch, and Ylla, *Animal Language* (London: Country Life, 1938, 12.

45. Huxley, "Darwin's Theory of Sexual Selection," 421, 431.

46. Ibid., 417.

47. Bartley, "Courtship and Continued Progress." Similar assumptions about "monogamy" in bird courtship and its parallel to human monogamy also defined the narrative of the recent Oscar-winning movie *March of the Penguins* (dir. Luc Jacquet, 2005).

48. Huxley, "Darwin's Theory of Sexual Selection," 417.

49. Bartley, "Courtship and Continued Progress."

50. Wells, Huxley, and Wells, *Science of Life*, vol. 2, 1239.

51. Marie Carmichael Stopes, *Married Love, or Love in Marriage*, with preface and notes by William J. Robinson, M.D. (New York: Critic and Guide Co., 1918), emphasis in original, available online at digital.library.upenn.edu/women/stopes/married/1918.html.

52. Peter Laipson, "'Kiss without Shame, for She Desires It': Sexual Foreplay in American Marital Advice Literature, 1900–1925," *Journal of Social History* 29, no. 3 (1996): 510.

53. In particular, Huxley worked closely with A. J. Marshall on sexual selection in Australian bower birds. In chapter 4, I discuss Marshall's research in conjunction with British ethology in the 1950s and 1960s.

54. William B. Provine, *Origins of Theoretical Population Genetics* (Chicago: University of Chicago Press, 1971; reprint, 2001); William B. Provine, "The Role of Mathematical Population Geneticists in the Evolutionary Synthesis of the 1930s and 40s," *Studies in the History of Biology* 2 (1978).

55. Joan Fisher Box, *R. A. Fisher: The Life of a Scientist* (New York: John Wiley and Sons, 1978). Fisher's research intersected several fields that today we consider quite distinct, making it difficult to label him as part of a single research community. For example, separate historiographic traditions have hailed Fisher as one of the "fathers" of both statistics and evolutionary theory.

56. Ronald Aylmer Fisher, *The Genetical Theory of Natural Selection* (Oxford: Clarendon Press, 1930).

57. Ronald Aylmer Fisher, "Some Hopes of a Eugenist," *Eugenics Review* 5 (1914); Ronald Aylmer Fisher, "The Evolution of Sexual Preference," *Eugenics Review* 7 (1915); Ronald Aylmer Fisher, "Positive Eugenics," *Eugenics Review* 9 (1917).

58. Bartley, "Conflicts in Human Progress."

59. Mary Bartley, "A Century of Debate: The History of Sexual Selection Theory (1871–1971)" (Ph.D. diss., Cornell University, 1994), 160.

60. Fisher, "Some Hopes," 309.

61. Fisher's framing of evolution as the quantity and quality of offspring being part of the same evolutionary process differs from that of biologists of Darwin's generation, who saw

the mechanisms of evolution as "sexual selection by which the fit are born, and natural selection by which the fittest survive." Alfred Russel Wallace, citing Hiram Stanley, in Wallace, "Human Selection," *Fortnightly Review* 48 (1890), 328.

62. Fisher, "Evolution of Sexual Preference," 189, 191.

63. Fisher, "Positive Eugenics," 206.

64. Fisher, "Evolution of Sexual Preference," 188.

65. I refer here to the *Variorum* edition of *Genetical Theory*. Ronald Aylmer Fisher, *The Genetical Theory of Natural Selection: A Complete Variorum Edition*, ed. J. H. Bennett (New York: Oxford University Press, 1999), x.

66. Ibid., 136.

67. Ibid., 137.

68. Ibid., 162.

69. "The individual is sacrificed, but the group bearing the hereditary characteristics of that individual has an increased chance of survival. The same argument applies with very little modification to the survival of heroism in closely knit, warring tribal societies . . . It is the heroes kindred who increase at the expense of less heroic strains." Fisher, Undated handwritten manuscript, Fisher Papers, Series 12/1: Eugenics, MSS 0013, AU-BSL.

70. William D. Hamilton, "The Evolution of Altruistic Behavior," *American Naturalist* 7 (1963); William D. Hamilton, "The Genetical Evolution of Social Behavior. I," *Journal of Theoretical Biology* 7 (1964); William D. Hamilton, "The Genetical Evolution of Social Behaviour. II," *Journal of Theoretical Biology* 7 (1964); Mary Jane West-Eberhard, "The Evolution of Social Behavior by Kin Selection," *Quarterly Review of Biology* 50, no. 1 (1975).

71. Bartley, "Century of Debate," 176–77.

72. Fisher, *Genetical Theory Variorum*, 255.

73. Ibid., 252.

74. Robinson's book went through fifteen editions between 1917 and 1939; I refer here to the 1939 edition. William J. Robinson, *Woman: Her Sex and Love Life*, 15th ed. (New York: Eugenics Publishing Co., 1939).

75. Ibid., 73.

76. Paul Popenoe, "Mate Selection," *American Sociological Review* 2, no. 5 (1937): 738.

77. Fisher to Haldane, May 1, 1931, and Fisher to Muller, July 7, 1930, both in Fisher Papers, Series I: Correspondence, MSS 0013, AU-BSL.

78. This explanation is tied up in debates between theoretical mathematical biologists and organismal biologists, much later in the century, concerning who was to blame for the "eclipse" of interest in sexual selection. I discuss the importance of these dynamics in writing the history of sexual selection in chapter 6. Egbert G. Leigh Jr., "Sex Ratio and Differential Mortality between the Sexes," *American Naturalist* 104, no. 936 (1970).

79. This reason for Fisher's ending his association with the Eugenics Society is noted by several individuals, including Joan Fisher Box in her biography of her father, Donald MacKenzie in his book on the history of British statistical theory, and Mary Bartley. Box, *R. A. Fisher*; Donald A. MacKenzie, *Statistics in Britain 1865–1930: The Social Construction of Scientific Knowledge* (Edinburgh: Edinburgh University Press, 1981), 183–213; Bartley, "Conflicts in Human Progress," 196.

80. Bartley, "Century of Debate," 192; Box, *R. A. Fisher*, 334.

81. A. J. Bateman, "Intra-sexual Selection in *Drosophila*," *Heredity* 2 (1948).

82. Oren Solomon Harman, *The Man Who Invented the Chromosome: A Life of Cyril Darlington* (Cambridge, MA: Harvard University Press, 2004).

83. A. J. Bateman to Fisher, May 10, 1948, Fisher Papers, Series 1: Correspondence, MSS 0013, AU-BSL.

84. Fisher to A. J. Bateman, May 12, 1948, Fisher Papers, Series 1: Correspondence, MSS 0013, AU-BSL. Fisher's statement proved prescient, because later drosophilists used Bateman's clear presentation of the general principle to argue for female-choice rather than male-choice tests of reproductive isolation. See also Robert L. Trivers, "Parental Investment and Sexual Selection," in *Sexual Selection and the Descent of Man, 1871–1971*, ed. Bernard Clark Campbell (Chicago: Aldine Publishing Co., 1972).

85. For these arguments as applied to human reproduction in the late nineteenth century, see Russett, *Sexual Science.*

86. This story was related to Fisher in two letters, one from E. B. "Henry" Ford and one from Julian Huxley. Ford to Fisher, Sept. 1931 (the source of the quotation); Huxley to Fisher, Sept. 29, 1931. In a separate letter to Fisher, MacBride, the antagonist in question, said much the same thing: "Who should be the judges of evolution? Surely only those who are best acquainted with the facts on which the theory of evolution is based viz systematists, paleontologists and embryologists." In effect, not Fisher! MacBride to Fisher, Dec. 20, 1932. All letters in Fisher Papers, Series 1: Correspondence, MSS 0013, AU-BSL.

87. Angelique Richardson, *Love and Eugenics in the Late Nineteenth Century: Rational Reproduction and the New Woman* (New York: Oxford University Press, 2003); Diane B. Paul, *Controlling Human Heredity: 1865 to the Present* (Amherst: Humanities Books, 1998); Rensing, "Feminist Eugenics."

88. Kevles, *In the Name of Eugenics*, 170.

Chapter 3 · *Branching Out, Scaling Up*

Epigraph. William K. Gregory, "Address to the AMNH at Noble's Memorial," Dec. 19, 1940, Noble biographical file, Special Collections, American Museum of Natural History Research Library, New York (hereafter, AMNH–Special Collections).

1. Richard W. Burkhardt Jr., *Patterns of Behavior: Konrad Lorenz, Niko Tinbergen, and the Founding of Ethology* (Chicago: University of Chicago Press, 2005); Donald A. Dewsbury, "A Brief History of the Study of Animal Behavior in North America," *Perspectives in Ethology* 8 (1989); Gregg Mitman and Richard W. Burkhardt Jr., "Struggling for Identity: The Study of Animal Behavior in America, 1930–1945," in *The Expansion of American Biology*, ed. Keith Benson, Jane Maienschein, and Ronald Rainger (New Brunswick, NJ: Rutgers University Press, 1991).

2. Gregg Mitman, *The State of Nature: Ecology, Community, and American Social Thought, 1900–1950* (Chicago: University of Chicago Press, 1992).

3. Dewsbury, "Brief History of the Study of Animal Behavior."

4. "Transcript of the Evolutionary Synthesis Conference," May 23, 1974, Evening Session, p. 4, Ernst Mayr, Evolutionary Synthesis Papers, B/M451t, folder 1.7, Manuscript Collections, American Philosophical Society, Philadelphia (hereafter, APS).

5. William K. Gregory, "Gladwyn Kingsley Noble (1894–1940)," *Year Book of the American Philosophical Society* 5 (1941); Mitman and Burkhardt, "Struggling for Identity."

6. Gladwyn Kingsley Noble, "The Experimental Animal from the Naturalist's Point of View," *American Naturalist* 78 (1939): 113.

7. Charles W. Myers, "A History of Herpetology at the American Museum of Natural History," *Bulletin of the American Museum of Natural History* 252 (2000): 25–28, 40. For a more colorful exploration of Noble's personality, see Frank A. Beach's autobiographical essays: "Confessions of an Imposter," in *Pioneers in Neuroendocrinology II*, ed. J. Meites, B. T. Donovan, and S. M. McLann (New York: Plenum Press, 1978); and "Frank A. Beach," in *A History of Psychology in Autobiography*, vol. 6, ed. Gardner Lindzey (Englewood Cliffs, NJ: Prentice-Hall, 1974).

8. Myers, "History of Herpetology at the AMNH," 25–28.

9. Theodosius Dobzhansky, *Genetics and the Origin of Species* (New York: Columbia University Press, 1937); Ernst Mayr, *Systematics and the Origin of Species from the Viewpoint of a Zoologist* (New York: Columbia University Press, 1942); George Gaylord Simpson, *Tempo and Mode in Evolution* (New York: Columbia University Press, 1944).

10. "Notes for the Trustee's Meeting," n.d., but attached to letter of Apr. 28, 1933, Noble Papers, Departmental folder, AMNH-Herpetology.

11. Elizabeth Hanson, *Animal Attractions: Nature on Display in American Zoos* (Princeton, NJ: Princeton University Press, 2002); Nicholas Jardine, James Secord, and Emma Spary, eds., *Cultures of Natural History* (Cambridge: Cambridge University Press, 1996); Gregg Mitman, "When Nature 'Is' the Zoo: Vision and Power in the Art and Science of Natural History," *Osiris, 2nd Series* 11 (1996); Nigel Rothfels, *Savages and Beasts: The Birth of the Modern Zoo* (Baltimore: Johns Hopkins University Press, 2002).

12. Przibram's Institute for Experimental Biology was also known as the Vivarium or "Sorcerers' Institute"; it was burned in the Soviet bombardment of Vienna in 1945. "Arthur Zitrin, 1988," Noble Papers, folder: Arthur Zitrin, AMNH-Herpetology.

13. "AMNH Annual Reports to the Trustees," 1928–1933, Noble Papers, Departmental folder, AMNH-Herpetology.

14. For a detailed account of the construction costs of the laboratory, see the "Departmental" section of Noble's archived papers at AMNH-Herpetology. The financial records for institutional expenditures were also published as the "Annual Report of the Trustees for the Year" until 1938, when they stopped including annual departmental expenditures.

15. Noble, "The Laboratory of Experimental Biology of the American Museum of Natural History," sent to Mr. Burden and the other trustees, May 19, 1933, Noble Papers, folder: W. Douglas Burden, AMNH-Herpetology.

16. "Remarks by Dr. Arthur Zitrin at the Memorial Meeting for Dr. Frank A. Beach, August 31, 1988, Berkeley, California," copy in Noble Papers, folder: Arthur Zitrin, AMNH-Herpetology.

17. See letters dating between Feb. 13, 1936, and Jan. 11, 1941, Noble Papers, folder: AMNH Administration #VI E—Personnel: WPA, AMNH-Herpetology.

18. Mitman and Burkhardt, "Struggling for Identity," 175.

19. Noble, "Laboratory of Experimental Biology of the American Museum of Natural History," 7.

20. Noble, "History of the Laboratory of Experimental Biology, American Museum of Natural History," Noble Papers, Departmental folder, AMNH-Herpetology.

21. Noble to Frank A. Beach, Nov. 1, 1937, Departmental files: Experimental Biology, AMNH-Herpetology.

22. Gladwyn Kingsley Noble, "Review of *The Elements of General Zoology* by William Dakin," *Science, New Series* 65, no. 1690 (1927): 501.

23. Noble, "Experimental Animal."

24. Gregory, "Address to the AMNH at Noble's Memorial"; Gregory, "Gladwyn Kingsley Noble (1894–1940)"; William K. Gregory, "Gladwyn Kingsley Noble (September 20, 1894–December 9, 1940)," *Science, New Series* 93, no. 2401 (1941).

25. William K. Gregory, *Our Face from Fish to Man: A Portrait Gallery of Our Ancient Ancestors and Kinsfolk together with a Concise History of Our Best Features* (New York: G. P. Putnam's Sons, 1929).

26. Noble had difficulty obtaining funding for behavioral field research. At the time of his death, he had completed the laboratory work but not the field work for his observations of gull behavior, for example.

27. Ruth Crosby Noble, *The Nature of the Beast: A Popular Account of Animal Psychology from the Point of View of a Naturalist* (Garden City, NY: Doubleday, Doran and Co., 1945); Myers, "History of Herpetology at the AMNH." Gregg Mitman has written extensively on Noble's commitment to reaching a popular audience through film and radio; see, for example, Mitman's "Cinematic Nature: Hollywood Technology, Popular Culture, and the American Museum of Natural History," *Isis* 84 (1993); and *Reel Nature: America's Romance with Wildlife on Film* (Cambridge, MA: Harvard University Press, 1999).

28. Myers, "History of Herpetology at the AMNH," 12.

29. Application of the term "organismal biologist" to Noble was proposed by Mitman and Burkhardt in 1991. Although the term did not come into widespread use until the 1960s, the intent with which it was used then—to specify a holistic approach to the biology of an organism—works equally well for Noble. Mitman and Burkhardt, "Struggling for Identity," 175.

30. Noble's work, in particular, mirrors similar work by Dobzhansky to demonstrate to naturalists that genetic studies of populations could inform their studies of species. Dobzhansky, *Genetics and the Origin of Species*.

31. This need for field expeditions to gather animals was also a problem for many contemporary field stations. Robert Kohler, "Place and Practice in Field Biology," *History of Science* 40 (2002).

32. Noble to Dr. Roy Chapman Andrews on the Department of Experimental Biology's Annual Report 1938, Jan. 14, 1939, Noble Papers, folder: Experimental Biology, AMNH-Herpetology.

33. Bernard Greenberg and Gladwyn Kingsley Noble, "Social Behavior of the American Chameleon (*Anolis carolinensis voigt*)," *Physiological Zoology* 17 (1944): 393–94.

34. Animals of the species *Anolis carolinensis* are sometimes referred to as "American chameleons" because they are capable of varying the color of their skin (shades of green and brown). They are not related to true chameleons, however, and the rest of the chapter refers to them as "anoles."

35. Greenberg and Noble, "Social Behavior of the American Chameleon," 392.

36. Ibid., 428.

37. Ibid., 400.

38. Noble, "Copulatory Process," 3, Turtle Research manuscript, folder: Gladwyn Kingsley Noble II—Research: H1-7 Spotted Turtles, Noble Papers, AMNH-Herpetology; see also Noble, "Homosexuality," Turtle Research manuscript. This research was interrupted by Noble's death and was never finished or published.

39. Noble, "Homosexuality," Turtle Research manuscript.

40. Gladwyn Kingsley Noble, "Courtship and Sexual Selection of the Flicker (*Colaptes auratus luteus*)," *Auk* 53 (1936).

41. Ibid., 274.

42. Ibid., 280.

43. Gladwyn Kingsley Noble and William Vogt, "An Experimental Study of Sex Recognition in Birds," *Auk* 52 (1935): 271.

44. Mitman and Burkhardt, "Struggling for Identity."

45. Greenberg and Noble, "Social Behavior of the American Chameleon," 402.

46. Noble to Professor Robert M. Yerkes, CRPS, NRC [Chairman of the Committee for Research in Problems of Sex, National Research Council], Apr. 8, 1935, Noble Papers, NRC folder 1, AMNH-Herpetology.

47. Greenberg and Noble, "Social Behavior of the American Chameleon."

48. Noble to Professor Robert M. Yerkes, Chairman of the Committee for Research in Problems of Sex, National Research Council, July 17, 1936, Noble Papers, NRC folder 1, AMNH-Herpetology.

49. Noble, "Experimental Animal."

50. Burden to Trubee Davison [fellow trustee of the AMNH], Jan. 13, 1941, folder: Frank A. Beach, 1196.1, Department of Animal Behavior, Central Archives, AMNH–Special Collections.

51. Frank A. Beach and C. S. Ford, *Patterns of Sexual Behavior* (New York: Harper, 1951); Frank Ambrose Beach, *Hormones and Behavior: A Survey of Interrelationships between Endocrine Secretions and Patterns of Overt Response* (New York: Harper, 1948); Frank Ambrose Beach, ed., *Sex and Behavior* (New York: John Wiley and Sons, 1965).

52. Charles Bogert to Beach, Aug. 7, 1941, Charles M. Bogert Papers, folder: Frank A. Beach, AMNH-Herpetology.

53. Beach to Andrews, Feb. 1, 1941, 1196.1, Department of Animal Behavior, Central Archives, AMNH–Special Collections.

54. Burden to Mr. Trubee Davidson, Jan. 13, 1941, 1196.1, Department of Animal Behavior, Central Archives, AMNH–Special Collections. Burden was a long-time friend and advocate of Noble's work. He collaborated with Noble on several film projects and greatly admired his research at the AMNH. As a trustee with an excellent fund-raising record, Burden wielded considerable influence at the museum until he retired from the board in 1962. See Gladwyn Kingsley Noble Papers and Charles M. Bogert Papers, W. Douglas Burden folders, AMNH-Herpetology; Mitman, "Cinematic Nature"; Mitman, *Reel Nature*.

55. Frank Ambrose Beach, "The Snark Was a Boojum," *American Psychologist* 5 (1950).

56. Frank Ambrose Beach, "Brains and the Beast," *Natural History* 53, June (1947): 284; Frank Ambrose Beach, "Brains and the Beast II," *Natural History* 53, Sept. (1947).

57. Donald A. Dewsbury, "Frank Ambrose Beach: 1911–1988," *American Journal of Psychology* 102, no. 3 (1989); Joshua Levens, "Sex, Neurosis and Animal Behavior: The Emergence of American Psychobiology and the Research of W. Horsley Gantt and Frank A. Beach" (Ph.D. diss., Johns Hopkins University, 2005).

58. Levens, "Sex, Neurosis and Animal Behavior," 225–50.

59. Lester R. Aronson, "Penile Spines of the Domestic Cat—Their Endocrine Behavior Relations," *Anatomical Record* 157 (1967); Lester R. Aronson and Gladwyn Kingsley Noble, "The Sexual Behavior of the Anura. 6. The Mating Pattern of *Bufo americanus, Bufo fowleri*, and *Bufo terrestris*," *American Museum Novitiates* 1250 (1944); Lester R. Aronson, Ethel Tobach, Daniel S. Lehrman, and Jay S. Rosenblatt, eds., *Development and Evolution of Behavior: Essays in Memory of T. C. Schneirla* (San Francisco: W. H. Freeman and Co., 1970); Dewsbury, "Frank Ambrose Beach."

60. Memorandum, Aronson to Mr. Wayne M. Faunce, June 29, 1949, 1196.1, Department of Animal Behavior, Central Archives, AMNH–Special Collections.

61. E. Thomas Gilliard, *Birds of Paradise and Bower Birds* (London: Weidenfeld and Nicolson, 1969).

62. Charles Bogert, "Isolating Mechanisms in Toads of the *Bufo debilis* Group in Arizona and Western Mexico," *American Museum Novitiates* 2100 (1962).

63. Much of this research has been forgotten, however. See, for example, Jerry A. Coyne and H. Allen Orr, *Speciation* (Sunderland, MA: Sinauer Associates, 2004), 3.

64. Charles M. Breder to Robert Cushman Murphy, Apr. 25, 1946, and Murphy's reply, May 23, 1946, both in Charles Cushman Murphy Papers, B M957, folder: Charles M. Breder, APS.

65. Myron Gordon, "The Genetics of a Viviparous Top-minnow *Platypoecilus*: The Inheritance of Two Kinds of Melanophores," *Genetics* 12 (1926); Myron Gordon and Donn Eric Rosen, "Genetics of Species Differences in the Morphology of the Male Genitalia of Xiphophoran Fishes," *Bulletin of the American Museum of Natural History* 95, no. 7 (1951).

66. C. M. Breder Jr., "An Experimental Study of the Reproductive Habits and Life History of the Cichlid Fish, *Aequidens latifrons* (Steindachner)," *Zoologica* 18, no. 1 (1934); C. M. Breder Jr. and C. W. Coates, "Sex Recognition in the Guppy, *Lebistes reticulatus* Peters," *Zoologica* 19 (1935); Charles M. Breder and Donn Eric Rosen, *Modes of Reproduction in Fishes* (Garden City, NY: Natural History Press, 1966).

67. After leaving the AMNH, Eugenie Clark quickly entered the public limelight as a beautiful female ichthyologist and authority on sharks, popularly known as "Shark Lady." Her autobiographical book, *Lady and a Spear* (1953), was a Book-of-the-Month Club selection, translated into seven languages, and encoded into Braille. Gary Kroll, *America's Ocean Wilderness: A Cultural History of Twentieth-Century Exploration* (Lawrence: University Press of Kansas, 2008), 124–51.

68. Philip Henry Gosse, *The Aquarium, an Unveiling of the Wonders of the Deep Sea* (London: J. Van Voorst, 1854).

69. Haskins and Haskins worked with three closely related species: *Lebistes reticulatus*

Peters, *Micropoecilia parae* Eigenmenn, and *Poecilia vivipara* Bloch and Schneider (according to 1949 taxonomic conventions). In their publication, the Haskins gave thanks to Dr. Dobzhansky "for suggesting the approach to the problem outlined herewith and for much critical help and encouragement, and to Dr. Myron Gordon, Dr. Lester Aronson, and Miss Eugenie Clark for reading the manuscript and offering most helpful suggestions and criticism." Caryl P. Haskins and Edna F. Haskins, "The Role of Sexual Selection as an Isolating Mechanism in Three Species of Poeciliid Fishes," *Evolution* 3, no. 2 (1949): 168.

70. Ibid., 166.

71. Ibid., 164.

72. Aronson to George W. Corner, Chairman of the Committee for Research in Problems of Sex, National Research Council, Mar. 12, 1949, Grantees: Aronson LR, NRC-CRPS Papers, 1946–52, National Research Council, Division of Medical Sciences, Committee for Research in Problems of Sex, 1920–1965, National Academies Archives, Washington, DC (hereafter, NRC-DIV-MED-CRPS).

73. Aronson to George W. Corner, Chairman of the Committee for Research in Problems of Sex, National Research Council, Mar. 13, 1950, Grantees: Aronson, LR, 1946–52, NRC-DIV-MED-CRPS.

74. Nikolaas Tinbergen, "Some Recent Studies of the Evolution of Sexual Behavior," in *Sex and Behavior*, ed. Frank A. Beach (New York: Robert E. Kreiger Publishing Co., 1974).

75. Aronson to George W. Corner, Chairman of the Committee for Research in Problems of Sex, National Research Council, Mar. 12, 1948, Grantees: Aronson, LR, 1946–52, NRC-DIV-MED-CRPS.

76. Eugenie Clark, Lester R. Aronson, and Myron Gordon, "Mating Behavior Patterns in Two Sympatric Species of Xiphophorin Fishes: Their Inheritance and Significance in Sexual Isolation," *Bulletin of the American Museum of Natural History* 103, art. 2 (1954).

77. The training Dobzhansky and Mayr received while in the Soviet Union and Germany, respectively, proved crucial to the methods and theoretical framework they transported to the United States. Mark Adams, "The Founding of Population Genetics: Contributions of the Chetverikov School, 1924–1934," *Journal of the History of Biology* 1 (1968); Mark Adams, "Towards a Synthesis: Population Genetics in Russian Evolutionary Thought," *Journal of the History of Biology* 3 (1970); Jürgen Haffer, *Ornithology, Evolution, and Philosophy: The Life and Science of Ernst Mayr, 1904–2005* (New York: Springer Verlag, 2007), 35–94; Ernst Mayr and William B. Provine, eds., *The Evolutionary Synthesis: Perspectives on the Unification of Biology* (Cambridge, MA: Harvard University Press, 1980); Vassiliki Betty Smocovitis, *Unifying Biology: The Evolutionary Synthesis and Evolutionary Biology* (Princeton, NJ: Princeton University Press, 1996).

78. Dobzhansky, *Genetics and the Origin of Species*.

79. Mayr, *Systematics and the Origin of Species*.

80. Although the synthesis architects later contended that their theories were valid for all animal and plant life, rampant hybridization in many plants makes defining a species on the basis of sexual incompatibility extremely difficult, as the research of Edgar Anderson vividly demonstrated. Botanists were finally brought into the synthesis community through G. Ledyard Stebbins. Edgar Anderson, *Introgressive Hybridization* (New York: John Wiley and Sons, 1949); Kim Kleinman, "His Own Synthesis: Corn, Edgar Anderson, and

Evolutionary Theory in the 1940s," *Journal of the History of Biology* 32 (1999); Vassiliki Betty Smocovitis, "G. Ledyard Stebbins and the Evolutionary Synthesis," *Annual Review of Genetics* 35 (2001); Vassiliki Betty Smocovitis, "Keeping Up with Dobzhansky: G. L. Stebbins, Plant Evolution and the Evolutionary Synthesis," *History and Philosophy of the Life Sciences* 28 (2006); G. Ledyard Stebbins, *Variation and Evolution in Plants* (New York: Columbia University Press, 1950).

81. Dobzhansky, *Genetics and the Origin of Species*, 312, quoting his earlier publication, "A Critique of the Species Concept in Biology," *Philosophy of Science* 2, no. 3 (1935): 354.

82. Mayr, *Systematics and the Origin of Species*, 120.

83. Dobzhansky, *Genetics and the Origin of Species*, 231–32.

84. Jason M. Baker, "Adaptive Speciation: The Role of Natural Selection in Mechanisms of Geographic and Non-geographic Speciation," *Studies in the History and Philosophy of Biology and the Biomedical Sciences* 36 (2005).

85. Theodosius Dobzhansky, "Experiments on Sexual Isolation in Drosophila. III. Geographic Strains of *Drosophila sturtevanti*," *Proceedings of the National Academy of Sciences of the United States of America* 30, no. 11 (1944). Dobzhansky and Mayr's experimental method was a quantified version of the method developed by Alfred Sturtevant more than a decade earlier. Alfred H. Sturtevant, "Experiments on Sex Recognition and the Problem of Sexual Selection in *Drosophila*," *Journal of Animal Behavior* 6 (1915).

86. See, for example, David Merrell, "Measurement of Sexual Isolation and Selective Mating," *Evolution* 4 (1950).

87. J. M. Rendel, "Genetics and Cytology of *Drosophila subobscura*. II. Normal and Selective Matings in *Drosophila subobscura*," *Journal of Genetics* 46 (1945).

88. Ibid., 299.

89. George Streisinger, "Experiments on Sexual Isolation in Drosophila. IX. Behavior of Males with Etherized Females," *Evolution* 2, no. 2 (1948).

90. A. J. Bateman, "Intra-sexual Selection in *Drosophila*," *Heredity* 2 (1948).

91. Donald A. Dewsbury, "The Darwin-Bateman Paradigm in Historical Context," *Integrative and Comparative Biology* 45, no. 5 (2005).

92. Bateman, "Intra-sexual Selection," 364.

93. Ibid., 362.

94. Ibid., 353.

95. Both Bateman and Rendel attributed the apparent choosiness of females to their natural "coyness" that only males of the appropriate species could hope to overcome. I discuss this again in chapter 6. Sarah Blaffer Hrdy, "Empathy, Polyandry, and the Myth of the Coy Female," in *Feminist Approaches to Science*, ed. Ruth Bleier (New York: Pergamon Press, 1986).

96. Bateman's and Streisinger's papers had significantly less influence on field research of the time.

97. One might aptly call Merrell "Bateman's Bulldog" for his enthusiastic attempts to convince all drosophilists of the importance of the female-choice experimental model in studying reproductive isolation. See David Merrell's, "Mating Preferences in Drosophila," *Evolution* 14 (1960); "Measurement of Sexual Isolation"; "Selective Mating as a Cause of Gene Frequency Changes in Laboratory Populations of *Drosophila melanogaster*," *Evolution*

7, no. 4 (1953); "Selective Mating in *Drosophila melanogaster*," *Genetics* 34 (1949); and "Sexual Isolation between *Drosophila persimilis* and *Drosophila pseudoobscura*," *American Naturalist* 88 (1954).

98. "Evolutionary Synthesis III, May 23, 1974, Evening Session," p. 4, Transcripts of the Evolutionary Synthesis Conference, B/M451t, folder 1.7 Genetics Discussion, APS. In this remark, Dobzhansky was referring to the Chetverikov school in Russia. As the subsequent conversation illustrates, such claims to mutations produced "*in nature*" were key to Chetverikov's identity as a "naturalist" and mirror Dobzhansky's identity as a naturalist.

99. Dobzhansky further contended that laboratories were unduly criticized by naturalists. "It is commonplace that the evolutionary changes which organisms underwent during the immense time span of the earth's history cannot be reproduced in laboratory experiments. However, too much stress has often been laid on this obvious limitation of the experimental method . . . Although such sceptical voices are now heard less and less often, it remains desirable to devise experiments which could serve as models of evolutionary changes that occur in natural populations." Theodosius Dobzhansky and Boris Spassky, "Evolutionary Changes in Laboratory Cultures of *Drosophila pseudoobscura*," *Evolution* 1, no. 3 (1947): 191.

100. Bogert, "Isolating Mechanisms"; E. Thomas Gilliard, "Feathered Dancers of Little Tobago [birds of paradise]," *National Geographic* 114, no. 9 (1958).

101. Although Aronson's fellow curators presumably approved of his early research, his later research became increasingly mechanistic and neurological, and the distance between his research programs and those of other scientists at the AMNH grew even wider. In the mid-1970s, when Aronson retired, the Department of Animal Behavior was disbanded permanently.

102. J. P. Scott, Theodore Schneirla (of the Department of Animal Behavior at the AMNH), and the entire Committee for the Study of Animal Societies under Natural Conditions (founded in 1946) agreed that "natural" behavior was best investigated with field studies. John Paul Scott, "Methodology and Techniques for the Study of Animal Societies," *Annals of the New York Academy of Sciences* 51 (1950).

Chapter 4 · Courtly Behavior

Epigraph. Nikolaas Tinbergen, "The Curious Behavior of the Stickleback," *Scientific American* 187, no. 5 (Dec. 1952): 22.

1. Colin Beer, "Ethology: The Zoologist's Approach to Behaviour, Part 2," *Tuatara* 12 (1964); Daniel S. Lehrman, "Ethology and Psychology," *Recent Advances in Biological Psychiatry* 4 (1962).

2. Whereas Tinbergen preferred to observe animal behavior in the field whenever possible, Lorenz preferred "the keeping of animals" as a way to more easily observe the full gamut of their behavioral repertoire. Richard W. Burkhardt Jr., "Ethology, Natural History, the Life Sciences, and the Problem of Place," *Journal of the History of Biology* 32 (1999): 501; Konrad Lorenz, "The Comparative Method in Studying Innate Behaviour Patterns," *Symposia of the Society of Experimental Biology* 4, Physiological Mechanisms in Animal Behaviour (1950): 235–36.

3. The group of American comparative psychologists that came into contact with Euro-

pean ethologists most often appears, in retrospect, to be those who self-identified as second-generation "psychobiologists," following somewhat in the tradition of Karl Lashley but without his absolute reductionism. They certainly saw their research as investigating questions at the boundary of zoology and psychology, and many of the later psychobiologists were funded by the National Research Council, Committee for Research on Problems of Sex (NRC-CRPS). Joshua Levens, "Sex, Neurosis and Animal Behavior: The Emergence of American Psychobiology and the Research of W. Horsley Gantt and Frank A. Beach" (Ph.D. diss., Johns Hopkins University, 2005); Nadine Weidman, "Psychobiology, Progressivism, and the Anti-progressive Tradition," *Journal of the History of Biology* 29 (1996).

4. Richard W. Burkhardt Jr., *Patterns of Behavior: Konrad Lorenz, Niko Tinbergen, and the Founding of Ethology* (Chicago: University of Chicago Press, 2005), 326–69.

5. Ibid., chap. 2, "British Field Studies of Behavior: Selous, Howard, Kirkman, and Huxley."

6. Burkhardt, "Ethology, Natural History"; Hans Kruuk, *Niko's Nature: The Life of Niko Tinbergen and His Science of Animal Behaviour* (New York: Oxford University Press, 2003).

7. The friendship of Lorenz and Tinbergen and its effect on their approaches to behavior is explored elegantly in Burkhardt, *Patterns of Behavior*.

8. Burkhardt, "Ethology, Natural History"; Burkhardt, *Patterns of Behavior*; Kruuk, *Niko's Nature*; Gregg Mitman, *Reel Nature: America's Romance with Wildlife on Film* (Cambridge, MA: Harvard University Press, 1999); Gregg Mitman and Richard W. Burkhardt Jr., "Struggling for Identity: The Study of Animal Behavior in America, 1930–1945," in *The Expansion of American Biology*, ed. Keith Benson, Jane Maienschein, and Ronald Rainger (New Brunswick, NJ: Rutgers University Press, 1991).

9. Lorenz, "Comparative Method," 238.

10. Colin Beer, "Ethology: The Zoologist's Approach to Behaviour, Part 1," *Tuatara* 11 (1963); Tinbergen, "Curious Behavior of the Stickleback," 25.

11. Beer, "Ethology, Part 1"; Beer, "Ethology, Part 2"; Burkhardt, *Patterns of Behavior*, 451–60, 373–83; Robert A. Hinde, "Consequences and Goals: Some Issues Raised by Dr. A. Kortland's Paper on 'Aspects and Prospects of the Concept of Instinct,'" *British Journal of Animal Behaviour* 3 (1956); Robert A. Hinde, "Ethological Models and the Concept of 'Drive,'" *British Journal for the Philosophy of Science* 6, no. 24 (1956); Robert A. Hinde, "Unitary Drives," *Animal Behaviour* 7 (1959).

12. Tinbergen, "Curious Behavior of the Stickleback," 26.

13. Charles Robert Darwin, *Descent of Man and Selection in Relation to Sex* (London: John Murray, 1871), vol. 1, 63; vol. 2, 69–71, 102, 112–13.

14. "Significantly, laboratory-reared or maintained specimens [of fruit flies] exhibit courtship behaviors that are quantitatively and qualitatively similar to those performed under natural field conditions." Herman T. Spieth, "Courtship Behavior in *Drosophila*," *Annual Review of Entomology* 19 (1974): 385.

15. Helen Spurway and J. B. S. Haldane, "The Comparative Ethology of Vertebrate Breathing. I. Breathing in Newts, with a General Survey," *Behaviour* 6 (1953): 8.

16. Although examples of ritualized, innate behaviors other than courtship were harder to find, these behaviors also attracted significant attention in the ethological community—for example, von Frisch's work on bee foraging and Lorenz's work on fixed-action patterns

in birds. Tania Munz, "The Bee Battles: Karl von Frisch, Adrian Wenner and the Honey Bee Dance Language Controversy," *Journal of the History of Biology* 38 (2005).

17. Robert A. Hinde, "Behavior and Speciation in Birds and Lower Vertebrates," *Biological Reviews* 34, no. 1 (1959).

18. Konrad Lorenz, *On Aggression*, trans. Marjorie Latzke (London: Methuen, 1966); Desmond Morris, *The Naked Ape: A Zoologist's Study of the Human Animal* (London: Jonathan Cape, 1967); Desmond Morris, *The Human Zoo* (New York: McGraw-Hill Book Co., 1969); Irenäus Eibl-Eibesfeldt, *Love and Hate: The Natural History of Behavior Patterns*, trans. Geoffrey Strachan (New York: Aldine de Gruyter, 1970).

19. For an American's rejection of this "typical" approach within comparative psychology, see Frank Ambrose Beach, "The Snark Was a Boojum," *American Psychologist* 5 (1950).

20. Tinbergen, "Animal Behaviour Research Group in the Department of Zoology, University of Oxford, 1949–1974," attached to a letter addressed to "Bill" [presumably Thorpe], Sept. 8, 1976, Tinbergen Papers, MS Eng., C.3125, A.4, Nikolaas Tinbergen, NCUACS 27.3.91, Bodleian Library, Department of Western Manuscripts, University of Oxford, Oxford (hereafter, OU-Bodleian).

21. Tinbergen to Noble, May 22, 1938, and Nov. 17, 1938, Noble Papers, folder: Tinbergen, AMNH-Herpetology.

22. See, for example, Gladwyn Kingsley Noble to Frank Fremont-Smith (Josiah Macy, Jr. Foundation), Feb. 15, 1937, and May 13, 1937, Noble Papers, folder: Foundations and Institutes—Josiah Macy, Jr., AMNH-Herpetology. "I would like to make it clear how our objectives differ significantly from those of other laboratories," Noble wrote. ". . . There are today many students of animal psychology practicing in the university laboratory. There are also some highly competent physiologists such as Bard, Crozier, and Rioch who are making important contributions to animal psychology. Neither group, however, is making any attempt to tap the enormous reservoir of accurate observations reported by field naturalists in all parts of the world. As I showed you we have abstracted over 10,000 papers in this field and our type-written notes fill eighteen feet of shelf space. This represents the work of an average number of eight translators working over a four-year period. These data cannot be waved aside as 'anecdotal' because many of these observers are exactly as analytical as Lorenz and all record what happens under natural conditions."

23. Noble to Tinbergen, Feb. 20, 1939, and Dec. 12, 1939, Noble Papers, folder: Tinbergen, AMNH-Herpetology.

24. Tinbergen continued: "I trust you that you would feel the same way when you were a citizen of a small free country like ours that was in danger of being overrun by a power which intends to bring you into a position of true slavery whenever it would get the slightest chance." Tinbergen to Noble, Feb. 14, 1940, Noble Papers, folder: Tinbergen, AMNH-Herpetology.

25. Konrad Lorenz, "Der Kumpan in der Umwelt des Vogels: der Artgenosse als auslösendes Moment sozialer Verhaltungsweisen," *Journal für Ornithologie* 83 (1935); Konrad Lorenz, "The Companion in the Bird's World," *Auk* 54 (1937).

26. Fremont Smith to Noble, May 13, 1937, Noble Papers, folder: Foundations and Institutes—Josiah Macy, Jr., AMNH-Herpetology. Fremont Smith sent the translation to (at least) the following people: Henry A. Murray Jr. (Harvard Psychological Clinic), Kurt Lewin

(State University of Iowa), George E. Coghill (Gainesville, Florida), Mrs. Gardener Murphy (Sarah Lawrence College), Stanley Cobb (Harvard Medical School, Neuropathology), Philip Bard (Johns Hopkins Medical School, Physiology), H. S. Liddell (Cornell University Medical College), W. Horsley Gantt (Phipps Psychiatric Clinic, Johns Hopkins University), Franz Alexander (Institute for Psychoanalysis, Chicago), William Malamud (State University of Iowa), Saul Rosenzweig (Worchester State Hospital), and Erik Homburger (Institute of Human Relations, Yale University).

27. Additional recipients of the translation probably included Glover Allen (Harvard University), William Burden, Eugene Odum (Huyck Preserve), David Davis (Biological Laboratory, Cambridge, MA), and Walter Clyde Allee (University of Chicago). Members of the Department of Experimental Biology at the AMNH—Frank Beach, Ted Schneirla, and Lester Aronson—certainly had access to a copy directly from Noble as well. Mitman and Burkhardt, "Struggling for Identity," n. 54.

28. Jay S. Rosenblatt, "Daniel Sanford Lehrman, June 1, 1919–August 27, 1972," *Biographical Memoirs of the National Academy of the Sciences* (1972): 232.

29. D. S. Lehrman, "A Critique of Konrad Lorenz's Theory of Instinctive Behavior," *Quarterly Review of Biology* 28, no. 4 (1953).

30. Rosenblatt, "D. S. Lehrman," 231; author's interview with Ethel Tobach, Feb. 25, 2004.

31. Burkhardt, *Patterns of Behavior.*

32. T. C. Schneirla, "Behavioral Development and Comparative Psychology," *Quarterly Review of Biology* 41, no. 3 (1966); Burkhardt, *Patterns of Behavior*, 362–68, 384–90.

33. Lehrman, "Critique of Konrad Lorenz," 353; T. C. Schneirla, "Levels in the Psychological Capacities of Animals," in *Philosophy for the Future: The Quest of Modern Materialism*, ed. Roy Wood Sellars, V. J. McGill, and Marvin Farber (New York: Macmillan Co., 1949).

34. Historian Donald Dewsbury notes that at least eighty-four Americans traveled to Europe to study ethology between 1945 and 1975, and at least fifteen ethologically trained Europeans moved to North America during the same period. This physical exchange of people, he argued, is at the core of ethology's success in the United States following the Second World War. Donald A. Dewsbury, "Americans in Europe: The Role of Travel in the Spread of European Ethology after World War II," *Animal Behaviour* 49 (1995).

35. Aubrey Manning, "The Ontogeny of an Ethologist," in *Studying Animal Behavior: Autobiographies of the Founders*, ed. Donald A. Dewsbury (Chicago: University of Chicago Press, 1989), 291.

36. For more on the importance of this conference and a detailed description of what occurred, see Burkhardt, *Patterns of Behavior*, 398–403; Paul E. Griffiths, "Instinct in the '50s: The British Reception of Konrad Lorenz's Theory of Instinctive Behaviour," *Biology and Philosophy* 4 (2004); and Kruuk, *Niko's Nature*. Also, see the published proceedings of the conference (complete with commentary by the participants), *Conference on Group Processes: Transactions* (New York: Josiah Macy Jr. Foundation Publishing, 1954).

37. Tinbergen, "3rd Extra Lecture on Behaviour, Hilary 1955; Ontogeny of Behaviour," Tinbergen Papers, MS Eng. C.3131, C.9, Nikolaas Tinbergen, NCUACS 27.3.91, OU-Bodleian.

38. Ibid. See also Tinbergen, "Lectures Animal Behaviour, 1957–8; 9 Ontogeny," Tinbergen Papers, MS Eng. C.3132, C.26, Nikolaas Tinbergen, NCUACS 27.3.91, OU-Bodleian.

39. Irenäus Eibl-Eibesfeldt and Sol Kramer, "Ethology: The Comparative Study of Animal Behavior," *Quarterly Review of Biology* 33, no. 3 (1958); Robert A. Hinde, *Animal Behaviour* (New York: McGraw-Hill Book Co., 1966). Eibl-Eibesfeldt's paper also served as a rebuttal to some of Lehrman's critiques. Although Tinbergen never published his own response, one can tell from the acknowledgments that both he and Konrad Lorenz offered comments on a draft of Eibl-Eibesfeldt's paper.

40. Nikolaas Tinbergen, "On Aims and Methods of Ethology," *Zeitschrift für Tierpsychologie* 20 (1963).

41. Beer, "Ethology, Part 2"; Burkhardt, "Ethology, Natural history"; Gregory Radick, "Primate Language and the Playback Experiment, in 1890 and 1980," *Journal of the History of Biology* 38 (2005); Gregory Radick, *Simian Tongue: The Long Debate about Animal Language* (Chicago: University of Chicago Press, 2007).

42. Tinbergen, "Aggression and Fear in the Normal Sexual Behaviour of Some Animals," Paris—Feb. 1961, Tinbergen Papers, MS Eng. C.3136, C.88, Nikolaas Tinbergen, NCUACS 27.3.91, OU-Bodleian; Beer, "Ethology, Part 1," 173.

43. Huxley's comments noted in Tinbergen, "Aggression and Fear in the Normal Sexual Behaviour of Some Animals."

44. For earlier discussions of ritualization in the ethological literature, see A. David Blest, "The Concept of 'Ritualisation,'" in *Current Problems in Animal Behaviour*, ed. William Homan Thorpe and Oliver Louis Zangwill (Cambridge: Cambridge University Press, 1961); Nikolaas Tinbergen, "'Derived' Activities: Their Causation, Biological Significance, Origin, and Emancipation during Evolution," *Quarterly Review of Biology* 27, no. 1 (1952).

45. "Properly to understand a piece of behaviour we have to appreciate its function in the life of the animal and its position in the whole behavioural repertoire of the animal. This necessitates study of the animal in its natural situation or in conditions that do not disguise the biological relevance of its behaviour." Beer, "Ethology, Part 1," 171.

46. John K. Kaufmann and Arleen Kaufmann, "Some Comments on the Relationship between Field and Laboratory Studies of Behaviour, with Special Reference to *Coatis*," *Animal Behaviour* 11, no. 4 (1963): 464.

47. Burkhardt, "Ethology, Natural History," 501; Lorenz, "Comparative Method," 234–35.

48. Again, see Burkhardt's extensive discussion of this side to ethological research. Burkhardt, *Patterns of Behavior*, 384–403.

49. W. H. Thorpe, "Comparative Psychology," *Annual Review of Psychology* 12 (1961): 28.

50. Female bower birds observed the finished bowers of the males and then chose a mate based on his behavioral and architectural display.

51. Jane Marshall, *Jock Marshall: One Armed Warrior*, Australian Science Archives Project on ASAPWeb, 1998, available online at www.asap.unimelb.edu.au/bsparcs/exhib/marshall/marshall.htm.

52. Ibid., chap. 3, "From the Fringe to the Centre." Leaving New Hebrides, Marshall continued to New Guinea to collect specimens for the Sydney Museum. There are many accounts of these New Hebrides–New Guinea research trips, authored by Marshall, his research partner Tom Harrisson, and Baker. John Randall Baker, *Man and Animals in the New Hebrides* (London: Routledge, 1929); John Randall Baker, *Scientific Results of the Oxford*

University Expedition to the New Hebrides, 1933–34 (Oxford: Oxford University Exploration Club, 1951); Tom Harrisson, *Savage Civilization* (London: Victor Gollancz, 1937); Alan John Marshall, *The Men and Birds of Paradise: Journeys through Equatorial New Guinea* (London: William Heinemann, 1938).

53. J. Marshall, *One Armed Warrior*.

54. Norman Chaffer, "The Spotted and Satin Bower-Birds: A Comparison," *Emu* 44, no. 3 (1944): 179–80.

55. Ernst Mayr and E. Thomas Gilliard, "Birds of Central New Guinea," *Bulletin of the American Museum of Natural History* 103 (1954).

56. Alan John Marshall, *Bower Birds: Their Displays and Breeding Cycles, a Preliminary Statement* (Oxford: Clarendon Press, 1954).

57. On the centrality of F. H. A. Marshall to British reproductive sciences in the twentieth century, see Adele E. Clarke, "Reflections on the Reproductive Sciences in Agriculture in the UK and US, ca. 1900–2000+," *Studies in the History and Philosophy of Biology and the Biomedical Sciences* 38 (2007). On the connection between Baker's interest in reproductive physiology and his controversial publications on race, see Michael G. Kenny, "Racial Science in Social Context: John R. Baker on Eugenics, Race, and the Public Role of the Scientist," *Isis* 95 (2004). Initially, Jock Marshall felt deeply indebted to Baker, and in 1952 he asked Baker for his support in applying for the chair of the Department of Zoology at Reading University. In his referee's report to the university, Baker not only mentioned Marshall's divorce, effectively calling his moral character into question, but also questioned his training as a classical zoologist. This incident strained their professional friendship. J. Marshall, *One Armed Warrior*, chap. 14, "Bower Birds, Books and People."

58. Alan John Marshall, "The Bower-Bird," in *Pacific Service of the BBC*, BBC Radio, broadcast May 9, 1947, 5:15–5:30 a.m. GMT. A copy of the transcript can be found in Marshall Papers, Series 16, box 51, file 5, MS 7132, National Library of Australia, Canberra (hereafter, NLA).

59. A. J. Marshall, *Bower Birds*, v.

60. Alan John Marshall, "Bower Birds," *Scientific American* 194, no. 6 (1956): 52; A. J. Marshall, *Bower Birds*, 172, 87.

61. Juliette Huxley (Julian's wife) wrote to Jane Marshall, after Jock's death: "You ask me for any memory I have of Jock. There is one thing I shall never forget. He dropped in on us one evening when Francis was in Brazil, and I was very worried as no letter had come from him for some time. Jock took a lively interest in Francis and had been very good to him at Jan Mayen. Anyway, he said: 'Don't worry about that little bastard, old girl. He'll be all right.' This was so typical of Jock . . . He was a bright and shining spirit, with the gaiety of a warm heart." Juliette Huxley to Jane Marshall, after Jock's death, Oct. 8, 1969, Marshall Papers, Series 6, file 44, MS 7132, NLA.

62. Ibid.

63. Julian S. Huxley, "Darwin's Theory of Sexual Selection and the Data Subsumed by It, in the Light of Recent Research," *American Naturalist* 72 (1938): 418–22; Julian S. Huxley, "The Present Standing of the Theory of Sexual Selection," in *Evolution: Essays on Aspects of Evolutionary Theory*, ed. Gavin de Beer (London: Oxford University Press, 1938).

64. A. J. Marshall, *Bower Birds*, 69–71.

65. Review in *Daily Telegraph and Morning Post* (London), Dec. 4, 1954, 6; copy in Marshall Papers, Series 4, file 22, MS 7132, NLA.

66. Review by M.G.B., *Biology*, June (1955); copy in Marshall Papers, Series 15, file 3, MS 7132, NLA.

67. L. Harrison Matthews, *Nature* 175, no. 4457, Apr. 2, 1955; copy in Marshall Papers, Series 15, file 3, MS 7132, NLA.

68. E. Thomas Gilliard, *Birds of Paradise and Bower Birds* (London: Weidenfeld and Nicolson, 1969), 55.

69. Manning, "Ontogeny of an Ethologist," 291; Kruuk, *Niko's Nature.*

70. Robert Kohler, *Lords of the Fly:* Drosophila *Genetics and the Experimental Life* (Chicago: University of Chicago Press, 1994).

71. In 1948 Mayr founded the journal *Evolution*, and as it gained in popularity, he began to cement his reputation as a central figure in mid-century evolutionary theory. Joseph Allen Cain, "Common Problems and Cooperative Solutions: Organizational Activity in Evolutionary Studies," *Isis* 84 (1993); Joseph Allen Cain, "Ernst Mayr as Community Architect: Launching the Society for the Study of Evolution and the Journal *Evolution*," *Biology and Philosophy* 9 (1994); Ernst Mayr, *Animal Species and Evolution* (Cambridge, MA: Belknap Press of Harvard University Press, 1963); Ernst Mayr, *The Growth of Biological Thought: Diversity, Evolution, and Inheritance* (Cambridge, MA: Belknap Press of Harvard University Press, 1982); Ernst Mayr and William B. Provine, eds., *The Evolutionary Synthesis: Perspectives on the Unification of Biology* (Cambridge, MA: Harvard University Press, 1980); Vassiliki Betty Smocovitis, "Disciplining Evolutionary Biology: Ernst Mayr and the Founding of the Society for the Study of Evolution (1939–1950)," *Evolution* 49 (1994); Vassiliki Betty Smocovitis, "Organizing Evolution: Founding the Society for the Study of Evolution (1930–1950)," *Journal of the History of Biology* 27 (1994).

72. Mayr to Konrad Lorenz, Nov. 10, 1949, Ernst Mayr Papers, HUGFP 14.7, box 7, folder 326, General Correspondence, 1931–1952, Harvard University Archives, Cambridge, Massachusetts (hereafter, HU).

73. Mayr to Tinbergen, Dec. 20, 1945, Ernst Mayr Papers, HUGFP 14.7, box 4, folder 160, General Correspondence, 1931–1952, HU.

74. Ernst Mayr, "Experiments on Sexual Isolation in Drosophila. VI. Isolation between *Drosophila pseudoobscura* and *Drosophila persimilis* and Their Hybrids," *Proceedings of the National Academy of Sciences of the United States of America* 32, no. 3 (1946); Ernst Mayr, "The Role of Antennae in the Mating Behavior of Female Drosophila," *Evolution* 4, no. 2 (1950).

75. Mayr, "Role of Antennae," 153.

76. Mayr to Tinbergen, Sept. 12, 1949, Ernst Mayr Papers, HUGFP 14.7, box 6, folder 313, General Correspondence, 1931–1952, HU.

77. Matthew Cobb, "A Gene Mutation Which Changed Animal Behaviour: Margaret Bastock and the Yellow Fly," *Animal Behaviour* 74 (2007).

78. Tinbergen to Mayr, Mar. 19, 1957, Ernst Mayr Papers, HUGFP 14.17, folder: Tinbergen, Nikolaas 1955–1959, Konrad Lorenz and Nicolas Tinbergen, 1953–1982, HU.

79. Author's interview with Claudine Petit, Dec. 2003; see also Burkhardt, *Patterns of Behavior*, 334–37.

80. Manning, "Drosophila and the Evolution of Behavior," 128.

81. The three main ethological journals at this time were *Animal Behaviour* (Oxford), *Zeitschrift für Tierpsychologie* (Berlin), and *Behaviour* (Leiden, Netherlands). In comparison with the 33 papers on Drosophila in *Animal Behavior*, all three journals combined published only 12 papers on gull behavior, 13 on crabs, 11 on cats, and 11 on sticklebacks.

82. Margaret Bastock, "A Gene Mutation Which Changes a Behavior Pattern," *Evolution* 10, no. 4 (1956): 421.

83. Ibid.; Herman T. Spieth, "Mating Behavior within the Genus *Drosophila* (Diptera)," *Bulletin of the American Museum of Natural History* 99, no. 7 (1952). Bastock's argument here echoes the importance of female mating behavior as a matter of species identity, as I described in chapter 3. See also Hinde, "Behavior and Speciation in Birds and Lower Vertebrates."

84. Cobb, "Gene Mutation"; Arthur W. Ewing and Aubrey Manning, "The Evolution and Genetics of Insect Behaviour," *Annual Review of Entomology* 12 (1967).

85. Maynard Smith published two papers in a series on the genetics and cytology of *Drosophila subobscura* that resulted from a massive collaboration of predominantly London-based geneticists. "The Genetics and Cytology of *Drosophila subobscura*," parts I–XI, *Journal of Genetics*, 1945–1955, included the following: J. M. Clarke and John Maynard Smith, "XI. Hybrid Vigor and Longevity," *Journal of Genetics* 53 (1955); John Maynard Smith and S. Maynard Smith, "VIII. Heterozygosity, Viability, and Rate of Development," *Journal of Genetics* 52 (1954); and J. M. Rendel, "II. Normal and Selective Matings in *Drosophila subobscura*," *Journal of Genetics* 46 (1945).

86. The mating dance involves a series of rapid side-to-side movements by the female, which the male has to mimic—if he cannot keep up, she will not mate with him. John Maynard Smith, "Fertility, Mating Behavior, and Sexual Selection in *Drosophila subobscura*," *Journal of Genetics* 54 (1956): 276. Maynard Smith commented on male flies' lack of capacity in an interview with the author, Dec. 2003; his quip about human courtship is quoted in Paul Erickson, "The Politics of Game Theory: Mathematics and Cold War Culture, 1944–1984" (Ph.D. diss., University of Wisconsin, 2006), chap. 5, "The Cold War in Nature."

87. Maynard Smith, "Fertility, Mating Behavior, and Sexual Selection," 275, 276.

88. Aubrey Manning, "Antennae and Sexual Receptivity in *Drosophila melanogaster* Females," *Science, New Series* 158, no. 3797 (1967): 136.

89. Petit interview. A similar phenomenon of females' rejection of males was mentioned by John Maynard Smith (who also did not believe that forced copulations are possible in *Drosophila*) in an interview with the author (Dec. 2003) and by Herman T. Spieth in his "Sexual Behavior and Isolation in *Drosophila* I. The Mating Behavior of Species of the *willistoni* Group," *Evolution* 1 (1947).

90. Manning, "Antennae and Sexual Receptivity."

91. Tinbergen, "Animal Behaviour 1972 Lecture 13: Reproductive Behaviour," Tinbergen Papers, MS Eng. C.3133, C.41, Nikolaas Tinbergen, NCUACS 27.3.91, OU-Bodleian.

92. Ibid.

93. Konrad Lorenz, *King Solomon's Ring: New Light on Animal Ways*, trans. Marjorie Kerr Wilson (New York: Thomas Y. Crowell Co., 1952); Lorenz, *On Aggression*.

94. After the publication of E. O. Wilson's *Sociobiology* (more on Wilson in chapter 6), human ethology was received much more critically. If readers were generally sympathetic to

the idea that humans had an innate aggressive drive, they were much less willing to consider that all (or even most) human behavior could similarly be reduced to biological terms. Peter M. Driver, "Toward an Ethology of Human Conflict" (review of *On Aggression* by Konrad Lorenz, *African Genesis* by Robert Ardrey, *The Territorial Imperative* by Robert Ardrey, and *Human Behavior: A New Approach* by Claire Russell and W. M. S. Russell), *Journal of Conflict Resolution* 11, no. 3 (1967); Irenäus Eibl-Eibesfeldt, "Human Ethology: Concepts and Implications for the Sciences of Man" (including peer commentary and the author's response), *Behavioral and Brain Sciences* 2, no. 1 (1979); Ullica Segerstråle, *Defenders of the Truth: The Battle for Science in the Sociobiology Debate and Beyond* (Oxford: Oxford University Press, 2002); Edward Osborne Wilson, *Sociobiology: The New Synthesis* (Cambridge, MA: Belknap Press of Harvard University Press, 1975).

95. Nikolaas Tinbergen, "On War and Peace in Animals and Men," *Science* 160, no. June 28 (1968).

96. Morris, *Naked Ape*. Social scientists' growing irritation with a zoological approach to the study of human behavior finally exploded with the publication of E. O. Wilson's *Sociobiology* in 1975.

97. Donna Haraway, *Primate Visions: Gender, Race, and Nature in the World of Modern Science* (New York: Routledge, 1989), pt. 2, "Decolonization and Multinational Primatology."

98. Several ant species were known to wage wars of annihilation on other ant species and to steal the eggs of other species to raise them as slaves, but among mammals, humans seemed unique in their capacity to kill members of their own species. Charlotte Sleigh, *Six Legs Better: A Cultural History of Myrmecology*, Animals, History, Culture series, ed. Harriet Ritvo (Baltimore: Johns Hopkins University Press, 2007).

99. In epimeletic behavior, healthy individuals of a species aid other individuals that are sick or injured. Lionel Tiger and Robin Fox, "The Zoological Perspective in Social Sciences," *Man, New Series* 1, no. 1 (1966): 80.

100. Eibl-Eibesfeldt, "Human Ethology"; Eibl-Eibesfeldt, *Love and Hate*; Irenäus Eibl-Eibesfeldt and Hans Hass, "Film Studies in Human Ethology," *Current Anthropology* 8, no. 5, pt. 1 (1967).

101. Eibl-Eibesfeldt's fellow human ethologist, Hans Hass, agreed that social behavior must be recorded unobserved if it is to be of use to ethologists. Hans Hass, *The Human Animal: The Mystery of Man's Behavior*, trans. J. Maxwell Brownjohn (London: Hodder and Stoughton, 1970), 11.

102. Lorenz, *King Solomon's Ring*, chap. 13, "Ecce Homo!"

103. Blest, "Concept of 'Ritualisation.'"

104. Julian S. Huxley, "Introduction: A Discussion on Ritualization of Behaviour in Animals and Man," *Philosophical Transactions of the Royal Society of London, Series B Biological* 251, no. 772 (1966).

105. Lorenz, *On Aggression*, 236–75. As noted, some ant species were known to wage wars of annihilation on other ant species; see Sleigh, *Six Legs Better*. Jane Goodall's later observations of a four-year war of annihilation between groups of chimpanzees would change this perspective (between 1974 and 1978). Goodall, *The Chimpanzees of Gombe: Patterns of Behavior* (Cambridge: Harvard University Press, 1986).

106. Lorenz's version of this story changed over time. At first he compared wolves and

rabbits, only changing it to wolves and doves years later when the dove became iconic for peace. Burkhardt suggests that Lorenz adapted his stories to elicit the strongest possible reaction in his readers. Burkhardt, *Patterns of Behavior*, 452–56.

107. Lorenz, "Comparative Method," 229.

108. M. F. Ashley Montagu, ed., *Man and Aggression* (Oxford: Oxford University Press, 1968), xi–xix.

109. Tinbergen, "On War and Peace in Animals and Men," 1414.

110. Robert A. Hinde, "Nature of Aggression," *New Society* 9 (1967); Robert Ardrey, *African Genesis* (New York: Atheneum, 1961); Nadine Weidman, "Popularizing the Ancestry of Man: Ardrey, Dart, and the Killer Instinct" (unpublished manuscript); J. E. Havel, "Review" (untitled review of *African Genesis*), *American Anthropologist* 66, no. 2 (1964).

111. Eibl-Eibesfeldt, *Love and Hate*, 46–47, 51.

112. Hass, *Human Animal*, 75.

113. Irenäus Eibl-Eibesfeldt, "Ritual and Ritualization from a Biological Perspective," in *Human Ethology: Claims and Limits of a New Discipline*, ed. M. von Cranach, K. Foppa, W. Lepenies, and D. Ploog (Cambridge: Cambridge University Press, 1979).

114. Morris, *Naked Ape*; Morris, *Human Zoo*.

115. The sexual habits of bonobos did not become a popular research topic until the 1980s. Bonobos' ubiquitous use of sex as a tool to diffuse social tension makes it unlikely for anyone today to claim humans as the "sexiest" ape. See, for example, Frans de Waal, *Bonobo: The Forgotten Ape* (Berkeley: University of California Press, 1998).

116. John Leonard, "The Baboons Do It Better," *New York Times*, Oct. 28, 1969.

117. George Gaylord Simpson, "What Is Man?" (review of *The Naked Ape*), *New York Times*, Feb. 4, 1968.

118. Fox concurrently lectured at the London School of Economics, and in 1967 he moved to Rutgers University, in New Jersey, to chair its new Anthropology Department.

119. Robin Fox, "The Evolution of Human Sexual Behavior," *New York Times Magazine*, Mar. 24, 1968, 79.

120. Ibid., 80–81.

121. Robin Fox, *Kinship and Marriage: An Anthropological Perspective* (New York: Penguin Books, 1967); Fox, "Evolution of Human Sexual Behavior," 93.

122. David Herlihy, "Biology and History: The Triumph of Monogamy," *Journal of Interdisciplinary History* 25, no. 4 (1995).

123. Jan Lindsten, ed., *Nobel Lectures, Physiology or Medicine, 1971–1980* (Singapore: World Scientific Publishing, Co., 1992), available online at http://nobelprize.org/nobel_prizes/medicine/laureates/1973/presentation-speech.html.

124. For a review of the early research on polygamy in birds, see Tim Birkhead, "Sperm Competition in Birds," *Trends in Ecology and Evolution* 2, no. 9 (1987).

Chapter 5 · A Science of Rare Males

Epigraph. Ernst Bösiger, "The Role of Sexual Selection in the Maintenance of the Genetical Heterogeneity of *Drosophila* Populations and Its Genetic Basis," in *Genetics of Behaviour*, ed. J. H. F van Abeelen (New York: Elsevier Publishing Company, 1974), 169.

1. Claudine Petit, "Le rôle de l'isolement sexuel dans l'evolution des population de *Drosophila melanogaster*," *Bulletin Biologique de la France et de la Belgique* 85 (1951); Claudine Petit, "L'isolement sexuel chez *Drosophila melanogaster* étude du mutant *white* et de son allélomorph *sauvage*," *Bulletin Biologique de la France et de la Belgique* 88 (1954); Claudine Petit, "Le déterminisme génétique et psycho-physiologique de la compétition sexuelle chez *Drosophila melanogaster*," *Bulletin Biologique de la France et de la Belgique* 92 (1958).

2. Robert Kohler, *Lords of the Fly:* Drosophila *Genetics and the Experimental Life* (Chicago: University of Chicago Press, 1994); William B. Provine, *Origins of Theoretical Population Genetics* (Chicago: University of Chicago Press, 1971; reprint, 2001).

3. Diane B. Paul, "'Our Load of Mutations' Revisited," *Journal of the History of Biology* 20, no. 3 (1987).

4. Charlotte Sleigh, *Six Legs Better: A Cultural History of Myrmecology*, Animals, History, Culture series, ed. Harriet Ritvo (Baltimore: Johns Hopkins University Press, 2007).

5. Richard M. Burian and Jean Gayon, "Genetics after World War II: The Laboratories at Gif (La génétique et les laboratoires de Gif)," *Cahiers pour l'histoire du CNRS* 7 (1990); Richard M. Burian and Jean Gayon, "The French School of Genetics: From Physiological and Population Genetics to Regulatory Molecular Genetics," *Annual Review of Genetics* 33 (1999); Jean Gayon and Richard M. Burian, "National Traditions and the Emergence of Genetics: The French Example," *Nature Reviews Genetics* 5, no. 2 (2004).

6. Petit's advisor, Georges Tessier, worked with Philippe L'Héritier on the evolution of large numbers of fruit flies in a "population cage." See, for example, P. L'Héritier and G. Tessier, "Une expérience de sélection naturelle: Courbe d'élimination du gène 'bar' dans une population de *Drosophiles* en équilibre," *Comptes Rendus des Séances de la Société de Biologie et de ses Filiales* 117 (1934). International recognition for a more physiological approach to French genetics came in 1965, when François Jacob, André Lwoff, and Jacques Monod won the Nobel Prize in Physiology or Medicine for their research on the genetic regulation of virus and protein synthesis.

7. The notable exception to this broad claim about French biologists is Ernst Bösiger, who also established strong connections to the American population genetics community through his association with Dobzhansky.

8. Yong-Kyu Kim, "Natural History of Lee Ehrman" (introduction to special issue), *Behavior Genetics* 35, no. 3 (2005).

9. Bösiger, "Role of Sexual Selection in Maintenance of Genetical Heterogeneity," 169.

10. Theodosius Dobzhansky, *Genetics and the Origin of Species* (New York: Columbia University Press, 1937); Ernst Mayr, *Systematics and the Origin of Species from the Viewpoint of a Zoologist* (New York: Columbia University Press, 1942); George Gaylord Simpson, *Tempo and Mode in Evolution* (New York: Columbia University Press, 1944); G. Ledyard Stebbins, *Variation and Evolution in Plants* (New York: Columbia University Press, 1950).

11. Noble as quoted in Henry Fairfield Osborn, "The Origin of Species. V. Speciation and Mutation," *American Naturalist* 61, no. 672 (1927): 16; and later in Ernst Mayr and William B. Provine, eds., *The Evolutionary Synthesis: Perspectives on the Unification of Biology* (Cambridge, MA: Harvard University Press, 1980), 125.

12. Osborn, "V. Speciation and Mutation," 40–41.

13. Mayr, *Systematics and the Origin of Species*.

14. Theodosius Dobzhansky, "Speciation as a Stage in Evolutionary Divergence," *American Naturalist* 74, no. 753 (1940).

15. Ernst Mayr, "Ecological Factors in Speciation," *Evolution* 1, no. 4 (1947): 270.

16. On the "pro" side, Mayr cited E. Stresemann, "Oekologische Sippen-, Rassen-, und Artunterschiede bei Vögeln," *Journal für Ornithologie* 91 (1942); and W. H. Thorpe, "The Evolutionary Significance of Habitat Selection," *Journal of Animal Ecology* 14 (1945). On the "anti" side, see Mayr, "Ecological Factors," 263.

17. Mayr and Provine, *Evolutionary Synthesis*, 20–22, 125. The early decades of the twentieth century have often be characterized as a period of Darwinian "eclipse," when few biologists sought to understand the process of evolution "for its own sake" (read: outside the eugenic application of genetics and evolution to human populations), instead relying on "non-Darwinian" mechanisms of organic diversity such as the inheritance of acquired characteristics (called "Lamarckism," after Jean-Baptiste Lamarck) and macro-mutations as a mechanism of evolutionary change (called "de Vriesian" evolution, after Hugo de Vries). For Thomas Hunt Morgan's advocacy of de Vries's mutation theory, and his dismissal of Lamarckism, see Thomas Hunt Morgan, *The Scientific Basis of Evolution* (London: Faber and Faber), 17–22, 188–202; and Garland Allen, "Hugo De Vries and the Reception of the 'Mutation Theory,'" *Journal of the History of Biology* 2, no. 1 (1969). For a historiographic review of the advantages and disadvantages of characterizing the early twentieth century as an "eclipse" of Darwinism, see Mark Largent, "The So-Called Eclipse of Darwinism," in *Descended from Darwin: Insights into the History of Evolutionary Studies, 1900–1970*, ed. Joe Cain and Michael Ruse (Philadelphia: American Philosophical Society, 2009).

18. Mayr, "Ecological Factors," 279.

19. Ernst Mayr, "Isolating Mechanisms," in *Vertebrate Speciation: A University of Texas Symposium*, ed. W. Frank Blair (Austin: University of Texas Press, 1961), 85.

20. H. J. Muller, "The Darwinian and Modern Conceptions of Natural Selection," *Proceedings of the American Philosophical Society* 93 (1949); Sewall Wright, "Evolution in Mendelian Populations," *Genetics* 16 (1931).

21. Susi Koref-Santibanez and C. H. Waddington, "The Origin of Sexual Isolation between Different Lines within a Species," *Evolution* 12, no. 4 (1958): 485.

22. Theodosius Dobzhansky, "Experiments on Sexual Isolation in Drosophila. III. Geographic Strains of *Drosophila sturtevanti*," *Proceedings of the National Academy of Sciences of the United States of America* 30, no. 11 (1944); Theodosius Dobzhansky and George Streisinger, "Experiments on Sexual Isolation in *Drosophila*. II. Geographic Strains of *Drosophila prosaltans*," *Proceedings of the National Academy of Sciences of the United States of America* 30, no. 11 (1944).

23. Dobzhansky and Streisinger, "II. Geographic Strains of *Drosophila prosaltans*," 345.

24. Dobzhansky, "III. Geographic Strains of *Drosophila sturtevanti*," 338–39.

25. Mayr to P. M. Sheppard, July 19, 1961, Ernst Mayr Papers HUGFP 74.7, box 8, folder 781, General Correspondence, 1952–1987, HU. The article in question is Ernst Mayr, "Changes in Genetic Environment and Evolution," in *Evolution as a Process*, ed. Julian S. Huxley, A. C. Hardy, and E. B. Ford (London: Allen and Unwin, 1954).

26. Suggestions that sympatric speciation might not be as unlikely as Mayr implied first cropped up among botanists. In 1955, for example, Richard M. Straw argued that sympatric

speciation in animal-pollinated plants would occur when a new plant hybrid was adopted by a new species-specific pollinator that maintained fidelity to that hybrid once established. Consistent "choice" on the part of the pollinator would lead to a new stable species of hybrid origin. Richard M. Straw, "Hybridization, Homogamy, and Sympatric Speciation," *Evolution* 9, no. 4 (1955).

27. See, for example, J. B. Gibson and J. M. Thoday, "Effects of Disruptive Selection. VI. A Second Chromosome Polymorphism," *Heredity* 17 (1962); J. M. Thoday, "Effects of Disruptive Selection: The Experimental Production of a Polymorphic Population," *Nature* 181, no. 4616 (1958); J. M. Thoday and J. B. Gibson, "Isolation by Disruptive Selection," *Nature* 193, no. 4821 (1962); J. M. Thoday and J. B. Gibson, "The Probability of Isolation by Disruptive Selection," *American Naturalist* 104 (1970).

28. Thoday and Gibson, "Isolation by Disruptive Selection."

29. For similar conclusions, see John Maynard Smith, "Disruptive Selection, Polymorphism and Sympatric Speciation," *Nature* 195 (1962): 62.

30. Thoday and Gibson, "Isolation by Disruptive Selection," 1166.

31. See, for example, Mark Ridley, *Evolution*, 3rd ed. (Oxford: Blackwell Publishing, 2003).

32. John Maynard Smith, "Sympatric Speciation," *American Naturalist* 100, no. 916 (1966).

33. Edmund Brisco Ford, *Ecological Genetics* (London: Methuen, 1964).

34. Maynard Smith, "Sympatric Speciation," 638.

35. Ibid., 637.

36. See, for example, G. L. Bush, "Modes of Animal Speciation," *Annual Review of Ecology and Systematics* 6 (1975); H. J. D. White, *Modes of Speciation* (San Francisco: W. H. Freeman and Co., 1978).

37. For a clear philosophical explication of the debate and the stakes involved (including concerns over the effects of radiation on future human evolution and eugenics), see John Beatty, "Weighing the Risks: Stalemate in the Classical/Balance Controversy," *Journal of the History of Biology* 29 (1987). For an explanation of the connection between the classical-balance debate and the "mutation load" of a population, see Michael Dietrich, "The Origins of the Neutral Theory of Molecular Evolution," *Journal of the History of Biology* 27, no. 1 (1994); Richard Lewontin, "Polymorphism and Heterosis: Old Wine in New Bottles and Vice Versa," *Journal of the History of Biology* 20, no. 3 (1987); and Paul, "'Our Load of Mutations' Revisited."

38. Hermann J. Muller, "Evolution by Mutation," *Bulletin of the American Mathematical Society* 64 (1958); Herman J. Muller, "Our Load of Mutations," *American Journal of Human Genetics* 2 (1950); Theodosius Dobzhansky, "A Review of Some Fundamental Concepts and Problems of Population Genetics," *Cold Spring Harbor Symposia on Quantitative Biology* 20 (1955).

39. Michelle Brattain, "Race, Racism, and Antiracism: UNESCO and the Politics of Presenting Science to the Postwar Public," *American Historical Review* 112, no. 5 (2007); M. F. Ashley Montagu, ed., *Statement on Race* (New York: Oxford University Press, 1972); William B. Provine, "Geneticists and Race," *American Zoologist* 26 (1986); UNESCO, ed., *The Race Question in Modern Science* (New York: UNESCO, 1956).

40. Theodosius Dobzhansky, *Mankind Evolving: The Evolution of the Human Species* (New Haven, CT: Yale University Press, 1962), 228.

41. For Muller's commitment to eugenics and the value of "superior" people, see Elof Axel Carlson, *Genes, Radiation, and Society: The Life and Work of H. J. Muller* (Ithaca, NY: Cornell University Press, 1981); Elof Axel Carlson, "The Parallel Lives of H. J. Muller and J. B. S. Haldane: Geneticists, Eugenicists, and Futurists," in *Haldane's Daedalus Revisited*, ed. Krishna R. Dronamraju (Oxford: Oxford University Press, 1995); Hermann Joseph Muller, *Out of the Night: A Biologist's View of the Future* (New York: Vanguard Press, 1935).

42. Dobzhansky, *Mankind Evolving*, 327–30.

43. Author's interview with Claudine Petit, Dec. 2003.

44. Julian S. Huxley, "Darwin's Theory of Sexual Selection and the Data Subsumed by It, in the Light of Recent Research," *American Naturalist* 72 (1938); Julian S. Huxley, "The Present Standing of the Theory of Sexual Selection," in *Evolution: Essays on Aspects of Evolutionary Theory*, ed. Gavin de Beer (London: Oxford University Press, 1938); Eliot B. Spiess, "Do Female Flies Choose Their Mates?" *American Naturalist* 119, no. 5 (1982).

45. Petit, "Isolement sexuel chez *D. melanogaster*"; Petit, "Déterminisme génétique"; Petit, "Rôle de l'isolement sexuel."

46. Eliot B. Spiess and Bozena Langer, "Mating Speed Control by Gene Arrangement Carriers in *Drosophila persimilis*," *Evolution* 18, no. 3 (1964); Eliot B. Spiess, Bozena Langer, and C. C. Li, "Chromosomal Adaptive Polymorphism in *Drosophila persimilis*. III. Mating Propensity of Homokaryotypes," *Evolution* 15 (1961).

47. Louis Levine, "Studies on Sexual Selection in Mice. I. Reproductive Competition between Albino and Black-Agouti Mice," *American Naturalist* 92, no. 862 (1958).

48. Ibid., 26.

49. Louis Levine's work was cited extensively after the mid-1970s as sexual selection theory took hold outside the *Drosophila* community. He published six more papers on sexual selection in mice between 1965 and 1982: Levine, "Sexual Selection in Mice. IV. Experimental Demonstration of Selective Fertilization," *American Naturalist* 101, no. 919 (1967); Levine, "Sexual Selection in Mice. VI. Effects of Fostering," *Journal of Heredity* 73, no. 5 (1982); Levine, C. A. Diakow, and G. E. Barsel, "Interstarin Fighting in Male Mice," *Animal Behavior* 13 (1965); Levine and Paul L. Krupa, "Studies on Sexual Selection in Mice. III. Effects of the Gene for Albinism," *American Naturalist* 100, no. 912 (1966); Levine and Barbara Lascher, "Studies on Sexual Selection in Mice. II. Reproductive Competition between Black and Brown Males," *American Naturalist* 99, no. 905 (1965); and Levine, Robert F. Rockwell, and Joseph Grossfield, "Sexual Selection in Mice. V. Reproductive Competition between +/+ and +/– males," *American Naturalist* 116, no. 1 (1980).

50. Ehrman received her Ph.D. from Columbia in 1959 and continued to work in Dobzhansky's laboratory until he left for the University of California, Davis, in 1970. Lee Ehrman, Boris Spassky, Olga Pavlovsky, and Theodosius Dobzhansky, "Sexual Selection, Geotaxis, and Chromosomal Polymorphism in Experimental Population of *Drosophila pseudoobscura*," *Evolution* 19, no. 1968 (1965).

51. Lee Ehrman's publications include "A Release Experiment Testing Mating Advantage of Rare *Drosophila* Males," *Behavioral Science* 15, no. 4 (1970); "Frequency Dependence of Mating Success in *Drosophila pseudoobscura*," *Genetic Research* 11 (1968); "Further Studies on

Genotype Frequency and Mating Success in *Drosophila*," *American Naturalist* 101, no. 921 (1967); "Genetics and Sexual Selection," in *Sexual Selection and the Descent of Man, 1871–1971*, ed. Bernard Clark Campbell (Chicago: Aldine Publishing Co., 1972); "Mating Success and Genotype Frequency in *Drosophila*," *Animal Behavior* 14 (1966); "Sensory Basis of Mate Selection in *Drosophila*," *Evolution* 23 (1969); "Simulation of the Mating Advantage in Mating of Rare *Drosophila* Males," *Science, New Series* 167 (1970); and "The Mating Advantage of Rare Males in *Drosophila*," *Proceedings of the National Academy of Sciences of the United States of America* 65, no. 2 (1970).

52. I can attest to both her hospitality and her culinary skill!

53. The idea that individuals with a greater amount of genetic variability were more likely to survive and produce greater numbers of offspring than individuals with less genetic variability was known variously as "hybrid vigor," the "heterotic effect," "heterosis," or "heterozygote advantage." One source of interest in this theory came from the remarkable success of hybrid corn in agriculture. See Kim Kleinman, "His Own Synthesis: Corn, Edgar Anderson, and Evolutionary Theory in the 1940s," *Journal of the History of Biology* 32, no. 2 (1999). Another source of interest derived from eugenic theories of selection. See, for example, Lewontin, "Polymorphism and Heterosis." For a contemporary spectrum of opinions on the topic, see the special issue "A Discussion of Hybrid Vigor," *Proceedings of the Royal Society of London, Series B Biological Sciences* 144, no. 915 (1955). The issue contains thirteen articles by luminaries in the field, such as Kenneth Mather, John Maynard Smith, Lionel Penrose, Cyril Darlington, Julian Huxley, and J. B. S. Haldane.

54. Bösiger, "Role of Sexual Selection," 172. I discussed Sturtevant's experiment in greater detail in chapter 2. Alfred Sturtevant, "Experiments on Sex Recognition and the Problem of Sexual Selection in *Drosophila*," *Journal of Animal Behavior* 6 (1915).

55. A. Faugeres, Claudine Petit, and E. Thibout, "The Components of Sexual Selection," *Evolution* 25, no. 2 (1971).

56. Julian Huxley, [Untitled review of *Mankind Evolving*], *Perspectives in Biology and Medicine*, 6 (1962): 144–48.

57. Huxley's first letter was to George Gaylord Simpson, Aug. 3, 1962, Julian Huxley Papers, MS 50, Series III, Correspondence, box 34, folder 3, Woodson Research Center, Fondren Library, Rice University, Houston, TX (hereafter, RU-Fondren).

58. Mayr to Huxley, Feb. 19, 1963, Julian Huxley Papers, MS 50, Series III, Correspondence, box 34, folder 2, RU-Fondren (emphasis added); Rensch to Huxley, Apr. 4, 1964, Julian Huxley Papers, MS 50, Series III, Correspondence, box 36, folder 4, RU-Fondren. This issue has been discussed extensively by Marc Swetlitz, who argues that "fitness in post–World War II evolutionary biology had a diversity of meanings, a condition that most biologists themselves recognized, although they did not endorse. Statements to the effect that fitness had one particular meaning or that geneticists had replaced a vague, value-laden definition by a rigorous, quantitative one were more prescriptions than descriptions of the situation." Marc Swetlitz, "The Contested Meaning and Usage of 'Fitness' in Post–World War II Evolutionary Biology" (unpublished ms., 1994), 36. On the multiplicity of meanings for the word *fitness* in modern biological discourse, see Costas B. Krimbas, "On Fitness," *Biology and Philosophy* 19 (2004).

59. Dobzhansky to Huxley, Mar. 14, 1963, Julian Huxley Papers, MS 50, Series III, Correspondence, box 34, folder 3, RU-Fondren.

60. Ken-Ichi Kojima, "Is There a Constant Fitness Value for a Given Genotype? No!" *Evolution* 25, no. 2 (1971).

61. D. Charlesworth and B. Charlesworth, "Sexual Selection and Polymorphism," *American Naturalist* 109, no. 968 (1975), 469; W. W. Anderson, "Polymorphism Resulting from the Mating Advantage of Rare Male Genotypes," *Proceedings of the National Academy of Sciences of the United States of America* 64 (1969).

62. Pomeroy Sinnock, "Frequency Dependence and Mating Behavior in *Tribolium castaneum*," *American Naturalist* 104, no. 939 (1970): 475.

63. D. Childress and I. C. McDonald, "Tests for Frequency-Dependent Mating Success in the House Fly," *Behavior Genetics* 3 (1973); Bruce Grant, G. Ann Snyder, and Steven F. Glessner, "Frequency-Dependent Mate Selection in *Mormoniella vitripennis*," *Evolution* 28, no. 2 (1974).

64. O'Donald used two computers, the Elliott 803 at the University College of North Wales and the Atlas I at the Science Research Council at Chilton, Berkshire. Peter O'Donald, "A General Model of Sexual and Natural Selection," *Heredity* 22 (1967): 499.

65. Peter O'Donald, "Natural Selection of Reproductive Rates and Breeding Times and Its Effect on Sexual Selection," *American Naturalist* 106, no. 949 (1972): 379.

66. The work on breeding of arctic skuas was published later as part of Peter O'Donald's *Genetic Models of Sexual Selection* (Cambridge: Cambridge University Press, 1980); and *The Arctic Skua* (Cambridge: Cambridge University Press, 1983).

67. Peter O'Donald, "Frequency Dependence and Polymorphism in Models of Sexual Selection," *American Naturalist* 111, no. 977 (1977); Peter O'Donald, "Mating Preferences and Sexual Selection in the Arctic Skua. II. Behavioral Mechanisms of the Mating Preferences," *Heredity* 39, no. 1 (1977); O'Donald, "Natural Selection of Reproductive Rates"; Peter O'Donald, "Polymorphisms Maintained by Sexual Selection in Monogamous Species of Birds," *Heredity* 32, no. 1 (1974).

68. O'Donald to Sheppard, June 10, 1968, Philip Sheppard Papers, MS Coll. 65, Series I, Correspondence, folder: Peter O'Donald, APS.

69. Peter O'Donald's publications at this time included: "A General Model of Mating Behaviour with Natural Selection and Female Preference," *Heredity* 40, no. 3 (1978); "Frequency Dependence"; "II. Behavioral Mechanisms of the Mating Preferences"; "Mating Preferences and Their Genetic Effects in Models of Sexual Selection for Colour Phases of the Arctic Skua," in *Population Genetics and Ecology*, ed. Samuel Karlin and Eviatar Nevo (New York: Academic Press, 1976); "Models of Sexual and Natural Selection in Polygamous Species," *Heredity* 31, no. 2 (1973); "Natural Selection of Reproductive Rates"; "Polymorphisms Maintained by Sexual Selection in Monogamous Species of Birds"; and "Theoretical Aspects of Sexual Selection: A Generalized Model of Mating Behaviour," *Theoretical Population Biology* 13, no. 2 (1978).

70. Peter O'Donald, "Do Female Flies Choose Their Mates? A Comment," *American Naturalist* 122, no. 3 (1983): 415.

71. O'Donald, "General Model of Sexual and Natural Selection," 499.

72. O'Donald, "Theoretical Aspects of Sexual Selection," 226; O'Donald, "General Model of Mating Behavior," 427–28.

73. O'Donald, "II. Behavioral Mechanisms of Mating Preferences," 111.

74. Author's interview with Lee Ehrman, Dec. 2002.

75. Wayne A. Hershberger and Maurice P. Smith, "Conditioning in *Drosophila melanogaster*," *Animal Behaviour* 15, no. 2–3 (1967).

76. Lucy B. Ellis and Seymour Kessler, "Differential Posteclosion Housing Experiences and Reproduction in *Drosophila*," *Animal Behaviour* 23, no. Part 4 (1975).

77. Petit interview; Ehrman interview; Mark B. Adams, ed., *The Evolution of Theodosius Dobzhansky: Essays on His Life and Thought in Russia and America* (Princeton, NJ: Princeton University Press, 1994).

78. Mark E. Borello, "The Rise, Fall and Resurrection of Group Selection," *Endeavour* 29, no. 1 (2005); Mark E. Borello, "Synthesis and Selection: Wynne Edwards Challenge to David Lack," *Journal of the History of Biology* 36 (2003): 531–66; George C. Williams, *Adaptation and Natural Selection: A Critique of Some Current Evolutionary Thought* (Princeton, NJ: Princeton University Press, 1966).

79. Michael Ruse, *Monad to Man: The Concept of Progress in Evolutionary Biology* (Cambridge, MA: Harvard University Press, 1997).

Chapter 6 · Selective History

Epigraph. Ernst Mayr to Julian Huxley, Oct. 3, 1967, Ernst Mayr Papers, HUGFP 14.17, HU.

1. In 1958, Sherwood Washburn and Virginia Avis identified "bipedal locomotion, the use of tools, and numerous functions of the brain, especially speech," as the characteristics distinguishing ape from human. S. L. Washburn and Virginia Avis, "Evolution of Human Behavior," in *Behavior and Evolution*, ed. Anne Roe and George Gaylord Simpson (New Haven, CT: Yale University Press, 1958), 421.

2. Jane Goodall, "Learning from the Chimpanzee: A Message Humans Can Understand," *Science* 282, no. 5397 (1998): 2184.

3. Goodall received her Ph.D. in ethology in 1965, under the direction of Robert Hinde. Marshall Flaum, dir., *Miss Goodall and the Wild Chimpanzees* (Washington, DC: National Geographic Society, 1965); Jane Goodall, *In the Shadow of Man* (Boston: Houghton Mifflin Books, 1971); Jane Goodall, "My Life with the Wild Chimpanzees," *National Geographic* 124, no. 8 (1963); Jane Goodall, "Tool-Using and Aimed Throwing in a Community of Free-Living Chimpanzees," *Nature* 201 (1964). As Haraway notes, the National Geographic Society quickly followed its success with one female primatologist by investing in others, especially Dian Fossey and Shirley Strum. Dian Fossey, "Making Friends with Mountain Gorillas," *National Geographic* 137, no. 1 (1970); Dian Fossey, "More Years with Mountain Gorillas," *National Geographic* 140, no. 10 (1971); Shirley Strum, "New Insights into Baboon Behavior," *National Geographic* 147, no. 5 (1975); Donna Haraway, *Primate Visions: Gender, Race, and Nature in the World of Modern Science* (New York: Routledge, 1989).

4. More than anything, this research implied that the barriers to spoken language in apes

were physical rather than mental. R. Allen Gardner and Beatrice T. Gardner, "Teaching Sign Language to a Chimpanzee," *Science, New Series* 165, no. 3894 (1969).

5. Gregg Mitman, *Reel Nature: America's Romance with Wildlife on Film* (Cambridge, MA: Harvard University Press, 1999), 177.

6. John Maynard Smith and George R. Price, "The Logic of Animal Conflict," *Nature* 246 (1973). Maynard Smith's first publication on game theory dates from 1972. In the introduction to the collection of essays in which this earlier piece was published, he wrote: "I would probably not have had the idea for this essay if I had not seen an unpublished manuscript on the evolution of fighting by Dr. George Price . . . Unfortunately, Dr. Price is better at having ideas than publishing them. The best I can do therefore is to acknowledge that if there is anything to the idea, the credit should go to Dr. Price and not to me." Maynard Smith's generosity as a thinker is as well-remembered as his inclination for meeting colleagues at the pub to hash out his biological theories. John Maynard Smith, "Game Theory and the Evolution of Fighting," in *On Evolution* (Edinburgh: Edinburgh University Press, 1972), vii–viii.

7. Richard Dawkins, *The Selfish Gene* (New York: Oxford University Press, 1976), 84; but see p. 287 of the revised (1989) edition for a later description of this characterization (of Maynard Smith and Price's paper) as "a bit over the top." Textbooks of evolution had already incorporated game theory by the 1980s. Paul Erickson, "The Politics of Game Theory: Mathematics and Cold War Culture, 1944–1984" (Ph.D. diss., University of Wisconsin, 2006), chap. 5, "The Cold War in Nature"; John Maynard Smith, "Game Theory and the Evolution of Behavior," *Proceedings of the Royal Society of London, Series B Biological Sciences* 205, no. 1161 (1979).

8. Dobzhansky to Simpson, Oct. 13, 1961, Dobzhansky Papers, B D65, Series 1 Correspondence, folder: GG Simpson, APS.

9. Mayr to Lorenz, May 20, 1964, Ernst Mayr Papers, HUGFP 14.17, folder: Konrad Lorenz 1963–1964, HU.

10. Edward Osborne Wilson, *Sociobiology: The New Synthesis*, 25th anniv. ed. (Cambridge, MA: Belknap Press of Harvard University Press, 2000), chap. 15, "Sex and Society"; Dawkins, *Selfish Gene*, chap. 9, "Battle of the Sexes"; Robert L. Trivers, "Parental Investment and Sexual Selection," in *Sexual Selection and the Descent of Man, 1871–1971*, ed. Bernard Clark Campbell (Chicago: Aldine Publishing Co., 1972).

11. Helena Cronin, *The Ant and the Peacock: Altruism and Sexual Selection from Darwin to Today* (Cambridge: Cambridge University Press, 1991), 231–49.

12. See, for example, Ruth Hubbard, "Have Only Men Evolved?" in *Women Look at Biology Looking at Women: A Collection of Feminist Critiques*, ed. Ruth Hubbard, Mary Sue Henifen, and Barbara Fried (Cambridge, UK: Schenkman Publishing Co., 1979).

13. John Maynard Smith repeats this suggestion in his foreword to Helena Cronin's *Ant and the Peacock*, ix–x.

14. Sarah Blaffer Hrdy, "Empathy, Polyandry, and the Myth of the Coy Female," in *Feminist Approaches to Science*, ed. Ruth Bleier (New York: Pergamon Press, 1986); Sarah Blaffer Hrdy and George C. Williams, "Behavioral Biology and the Double Standard," in *Social Behavior of Female Vertebrates* (New York: Academic Press, 1983).

15. Author's interview with Robert Trivers, Mar. 2004. Also, in his reciprocal altruism

paper, Trivers's citations on human behavior all come from the research of anthropologists involved in the production of MACOS—notably, John Marshall, Asen Balikci, and Irven DeVore. Robert L. Trivers, "The Evolution of Reciprocal Altruism," *Quarterly Review of Biology* 46 (1971).

16. Ullica Segerstråle, *Defenders of the Truth: The Battle for Science in the Sociobiology Debate and Beyond* (Oxford: Oxford University Press, 2002), 84.

17. Mayr to Mark Kirkpatrick, July 9, 1984, Ernst Mayr Papers, HUGFP 74.7, box 32, folder 1382, HU.

18. Robert L. Trivers, "Parent-Offspring Conflict," *American Zoologist* 14 (1974); Trivers, "Parental Investment and Sexual Selection"; Trivers, "Evolution of Reciprocal Altruism"; Robert L. Trivers and D. E. Willard, "Natural Selection of Parental Ability to Vary Sex-Ratio of Offspring," *Science* 179, no. 4068 (1973).

19. Robert L. Trivers, "A Full-Force Storm with Gale Winds Blowing: A Talk with Robert Trivers," Oct. 18, 2004, available online at www.edge.org/3rd_culture/trivers04/trivers04_index.html.

20. Trivers interview.

21. Bernard Clark Campbell, ed., *Sexual Selection and the Descent of Man, 1871–1971* (Chicago: Aldine Publishing Co., 1972).

22. Trivers interview.

23. Ibid.; Trivers, "Parental Investment and Sexual Selection"; A. J. Bateman, "Intrasexual Selection in *Drosophila*," *Heredity* 2 (1948): 349–68.

24. Citation analysis through the Science Citation Index online at http://isi3.isiknowledge.com.

25. Segerstråle notes that Trivers vigorously defended Wilson both in public and in print and took an instant dislike to Richard Lewontin, another of Dobzhansky's graduate students, who helped lead the attack against *Sociobiology*. Segerstråle, *Defenders of the Truth*, 82.

26. Trivers interview.

27. The number of articles published on the subjects of female choice and sexual selection increased linearly from the mid-1970s to the mid-1980s, when there was a huge jump in interest. This effect is due both to an increase in articles about sexual selection in long-standing journals such as *Animal Behaviour* and *Behaviour* and to the growing popularity of new specialized journals such as *Behavioral Ecology and Sociobiology* (first published in 1976) and *Ethology and Sociobiology* (1979).

28. William D. Hamilton, "The Genetical Evolution of Social Behaviour. I," *Journal of Theoretical Biology* 7 (1964); William D. Hamilton, "The Genetical Evolution of Social Behaviour. II," *Journal of Theoretical Biology* 7 (1964). See also Erickson, "Politics of Game Theory," chap. 5, "Cold War in Nature."

29. Trivers, "Parental Investment and Sexual Selection," 138, 144.

30. George C. Williams, *Adaptation and Natural Selection: A Critique of Some Current Evolutionary Thought* (Princeton, NJ: Princeton University Press, 1966).

31. The rare-male effect provides at least one exception to this generalization about group selection, however. V. C. Wynne-Edwards, "Self-Regulating Systems in Populations of Animals," *Science, New Series* 137, no. 3665 (1965).

32. Mark E. Borello, "The Rise, Fall and Resurrection of Group Selection," *Endeavour* 29, no. 1 (2005).

33. Williams, *Adaptation and Natural Selection*, 184–85. Williams identified Fisher as the original source of the idea.

34. D. Otte, "Historical Development of Sexual Selection Theory," in *Sexual Selection and Reproductive Competition in Insects*, ed. Murray S. Blum and Nancy A. Blum (New York: Academic Press, 1979).

35. A. Zahavi, "Mate Selection: A Selection for a Handicap," *Journal of Theoretical Biology* 53, no. 1 (1975).

36. William D. Hamilton and Marlene Zuk, "Heritable True Fitness and Bright Birds: A Role for Parasites?" *Science, New Series* 218, no. 4570 (1982); William D. Hamilton and Marlene Zuk, "Parasites and Sexual Selection," *Nature* 341 (1989).

37. Wilson, *Sociobiology*, 562.

38. I think of the situation as similar to the ways in which the publication of Charles Darwin's *On the Origin of Species* (1859) structured later reactions to evolutionary theory and erased from historical memory Robert Chamber's *Vestiges of the Natural History of Creation* (1844). On this topic, I highly recommend James Secord's *Victorian Sensation: The Extraordinary Publication, Reception, and Secret Authorship of Vestiges of the Natural History of Creation* (Chicago: University of Chicago Press, 2001). Historian Ullica Segerstråle has described the ensuing debates over sociobiology as a dramatic opera. Segerstråle, *Defenders of the Truth*, 6–9.

39. Alcock to Mayr, Apr. 28, 1980, Ernst Mayr Papers, General Correspondence, 1952–1987, HUGFP 74.7, box 28, folder 1318, HU.

40. Edward O. Wilson, *Naturalist* (Washington, DC: Island Press / Shearwater Books, 1994): 328–34. On Trivers, see Drake Bennett, "The Evolutionary Revolutionary," *Boston Globe*, Mar. 27, 2005. Some evolutionary psychologists look instead to Dick Alexander's application of evolutionary theory to human behavior as the true birth of the field. Richard D. Alexander, *Darwinism and Human Affairs* (Seattle: University of Washington Press, 1979).

41. Wilson, *Sociobiology*, chap. 15, "Sex and Society," 314.

42. The cover of the third edition—the text is the same as the second edition (1989), with a new introduction by Dawkins—declares it "THE MILLION COPY INTERNATIONAL BESTSELLER." Richard Dawkins, *The Selfish Gene: 30th Anniversary Edition* (Oxford: Oxford University Press, 2006).

43. Ibid., v, 2.

44. Williams, *Adaptation and Natural Selection*; John Maynard Smith, *The Theory of Evolution* (New York: Penguin Books, 1958); Maynard Smith and Price, "Logic of Animal Conflict"; William D. Hamilton, "Extraordinary Sex Ratios: A Sex-Ratio Theory for Sex Linkage and Inbreeding Has New Implications in Cytogenetics and Entomology," *Science, New Series* 156 (1967); Hamilton, "Evolution of Altruistic Behavior"; Hamilton, "Genetical Evolution of Social Behaviour. I"; Hamilton, "Genetical Evolution of Social Behaviour. II"; Trivers, "Parent-Offspring Conflict"; Trivers, "Parental Investment and Sexual Selection"; Trivers, "Evolution of Reciprocal Altruism."

45. Trivers interview.

46. Dawkins, *Selfish Gene*, 141–42. Dawkins's analysis of sex differences based on gamete physiology strongly echoed Geddes and Thompson's treatise on sex differences eighty-six years earlier. Patrick Geddes and J. Arthur Thompson, *The Evolution of Sex*, rev. ed. (London: Walter Scott, 1901), 117, 289–91.

47. Dawkins, *Selfish Gene*, 300–301.

48. Ibid., 165.

49. Dawkins, *Selfish Gene*; Segerstråle, *Defenders of the Truth*, 53.

50. Betty Friedan, *The Feminine Mystique* (New York: W. W. Norton, 1963), 15.

51. William H. Masters, Virginia E. Johnson, and the Reproductive Biology Research Foundation (St. Louis, Missouri), *Human Sexual Response* (Boston: Little, Brown, and Co., 1966). The FDA's declaration that oral contraceptives were safe for public consumption came in 1960.

52. See, for example, Anthony Leeds, Barbara Beckwith, Chuck Madansky, David Culver, Elizabeth Allen, Herb Schreier, Hiroshi Inouye, Jon Beckwith, Larry Miller, Margaret Duncan, Miriam Rosenthal, Reed Pyeritz, Richard C. Lewontin, Ruth Hubbard, Steven Chorover, and Stephen Jay Gould, "Against 'Sociobiology,'" *New York Review of Books* 22, no. 18 (1975); Conrad Hal Waddington, "Mindless Societies," *New York Review of Books* 22, no. 13 (1975); Edward O. Wilson, "For 'Sociobiology,'" *New York Review of Books* 22, no. 20 (1975).

53. Ruth Bleier, *Science and Gender: A Critique of Biology and Its Theories on Women* (New York: Pergamon Press, 1984), chap. 2, "Sociobiology, Biological Determinism, and Human Behavior"; Anne Fausto-Sterling, *Myths of Gender: Biological Theories about Men and Women* (New York: Basic Books, 1985); Donna Haraway, *Primate Visions: Gender, Race, and Nature in the World of Modern Science* (New York: Routledge, 1989); Hrdy and Williams, "Behavioral Biology and the Double Standard"; Hubbard, "Have Only Men Evolved?"; Marian Lowe and Ruth Hubbard, "Sociobiology and Biosociology: Can Science Prove the Biological Basis of Sex Differences in Behavior?" in *Genes and Gender II*, ed. Ruth Hubbard and Marian Lowe (New York: Gordian Press, 1979); J. Sayers, *Biological Politics: Feminist and Anti-Feminist Perspectives* (New York: Tavistock Publications, 1982).

54. See "Acknowledgements" in Wilson, *Sociobiology*.

55. Bateman, "Intra-Sexual Selection"; Hrdy, "Empathy, Polyandry, and Myth of the Coy Female," 119.

56. Hrdy, "Empathy, Polyandry, and Myth of the Coy Female."

57. Ibid., 135–39.

58. Citation analysis through the Science Citation Index online at http://isiknowledge .com indicates that, as of December 2008, Trivers's "Parental Investment and Sexual Selection" had been cited in more than five thousand publications since its publication in 1972. Interest in his chapter grew almost exponentially until 1980, at which time the number of citations continued to increase but in a more linear fashion.

59. Darwin, Fisher, Bateman, Merrell, Petit, Ehrman, Maynard Smith, O'Donald, Williams, and Hamilton had all used this argument as well.

60. Vassiliki Betty Smocovitis, "The 1959 Darwin Centennial Celebration in America," *Osiris, 2nd Series* 14, Commemorative Practices in Science: Historical Perspectives on the Politics of Collective Memory (1999).

61. Sol Tax, ed., *Evolution after Darwin*, vol. 1: *The Evolution of Life: Its Origin, History, and Future* (Chicago: University of Chicago Press, 1960); Sol Tax, ed., *Evolution after Darwin*, vol. 2: *The Evolution of Man: Man, Culture, and Society* (Chicago: University of Chicago Press, 1960); Sol Tax and Charles Callender, eds., *Evolution after Darwin*, vol. 3: *Issues in Evolution: The University of Chicago Centennial Discussions* (Chicago: University of Chicago Press, 1960).

62. Ghiselin's book won the History of Science Society's Pfizer Prize in 1969. Michael T. Ghiselin, *The Triumph of the Darwinian Method* (Berkeley: University of California Press, 1969).

63. Ernst Mayr and William Provine, eds., *The Evolutionary Synthesis: Perspectives on the Unification of Biology* (Cambridge, MA: Harvard University Press, 1980).

64. Pnina G. Abir-Am, "The Politics of Macromolecules: Molecular Biologists, Biochemists, and Rhetoric," *Osiris, 2nd Series* 7, Science after '40 (1992); Ron Amundson, *The Changing Role of the Embryo in Evolutionary Thought: Roots of Evo-Devo* (New York: Cambridge University Press, 2005); John Beatty, "Evolutionary Anti-reductionism: Historical Reflections," *Biology and Philosophy* 5 (1990); John Beatty, "The Proximate/Ultimate Distinction in the Multiple Careers of Ernst Mayr," *Biology and Philosophy* 9 (1994); Soraya de Chadarevian, *Designs for Life: Molecular Biology after World War II* (Cambridge: Cambridge University Press, 2002), especially chap. 8, "The Origins of Molecular Biology Revisited"; Donald A. Dewsbury, "On the Utility of the Proximate-Ultimate Distinction in the Study of Animal Behavior," *Ethology* 96 (1994); Michael R. Dietrich, "Paradox and Persuasion: Negotiating the Place of Molecular Evolution within Evolutionary Biology," *Journal of the History of Biology* 31 (1998); Philip Kitcher, "1953 and All That: A Tale of Two Sciences," *Philosophical Review* 93, no. 3 (1984); Michel Morange, *A History of Molecular Biology*, trans. Matthew Cobb (Cambridge, MA: Harvard University Press, 1998); Vassiliki Betty Smocovitis, *Unifying Biology: The Evolutionary Synthesis and Evolutionary Biology* (Princeton, NJ: Princeton University Press, 1996).

65. Beatty, "Proximate/Ultimate Distinction"; Ernst Mayr, "Cause and Effect in Biology," *Science* 134 (1961).

66. Ernst Mayr, *The Growth of Biological Thought: Diversity, Evolution, and Inheritance* (Cambridge, MA: Belknap Press of Harvard University Press, 1982).

67. Segerstråle, *Defenders of the Truth*, 85.

68. Charles Robert Darwin, *Descent of Man and Selection in Relation to Sex* (London: John Murray, 1871), 2 vols.; Ronald Aylmer Fisher, *The Genetical Theory of Natural Selection*, 2nd rev. ed. (New York: Dover Publications, 1958); Julian S. Huxley, "Darwin's Theory of Sexual Selection and the Data Subsumed by It, in the Light of Recent Research," *American Naturalist* 72 (1938); Julian S. Huxley, "The Present Standing of the Theory of Sexual Selection," in *Evolution: Essays on Aspects of Evolutionary Theory*, ed. Gavin de Beer (London: Oxford University Press, 1938).

69. Alfred Russel Wallace, "The Colors of Animals and Plants (part 1)," *American Naturalist* 11, no. 11 (1877); Alfred Russel Wallace, "The Colors of Animals and Plants (part 2)," *American Naturalist* 11, no. 12 (1877). For Trivers's "resurrection" of female choice, see Bartley, "Conflicts in Human Progress"; Bartley, "Courtship and Continued Progress"; Cronin, *Ant and the Peacock*; Simon J. Frankel, "The Eclipse of Sexual Selection Theory," in *Sexual*

Knowledge, Sexual Science: The History of Attitudes to Sexuality, ed. Roy Porter and M. Te- ich (Cambridge: Cambridge University Press, 1994); Geoffrey F. Miller, "How Mate Choice Shaped Human Nature: A Review of Sexual Selection and Human Evolution," in *Handbook of Evolutionary Psychology: Ideas, Issues, and Applications*, ed. C. Crawford and D. Krebs (Mah- wah, NJ: Lawrence Erlbaum, 1998).

70. Kenneth C. Waters and Albert Van Helden, eds., *Julian Huxley, Biologist and States- man of Science: Proceedings of a Conference Held at Rice University, 25–27 September 1987* (Houston, TX: Rice University Press, 1992).

71. Also, Mayr and Simpson were still alive in the early 1980s (Simpson died in 1984, Mayr in 2005; both Huxley and Dobzhansky died in 1975).

72. Robert K. Selander, "On Mating Systems and Sexual Selection," *American Naturalist* 99, no. 906 (1965), 129–30; Robert K. Selander, "Sexual Selection and Dimorphism in Birds," in Campbell, *Sexual Selection and the Descent of Man*, 223.

73. Two chapters in Campbell's volume focused primarily on birds, one on *Drosophila*, one on primates, and two on humans. It was, of course, a volume to commemorate Dar- win's *Descent of Man*, so an emphasis on sexual selection is not in itself surprising. All of the contributions to the volume, however, were review articles that summarized work from the previous decades.

74. Lee Ehrman, "Genetics and Sexual Selection"; Theodosius Dobzhansky, "Genetics and the Races of Man"; and Ernst Mayr, "Sexual Selection and Natural Selection," all in Campbell, *Sexual Selection and the Descent of Man, 1871–1971*.

75. Dobzhansky's chapter, however, addressed intraspecific female choice in human races, so that cannot have been the sole reason for the lack of attention these chapters received.

76. Gordon Allen, "Sexual Selection and the Descent of Man, 1871–1971—Campbell B," *Social Biology* 20, no. 3 (1973).

77. Selander, "Sexual Selection and Dimorphism in Birds," 181; Ehrman, "Genetics and Sexual Selection," 124, 126; Ernst Caspari, "Sexual Selection in Human Evolution," in Camp- bell, *Sexual Selection in Descent of Man, 1871–1971*, 342; Mayr, "Sexual Selection and Natural Selection," 100. Mayr's definition had less to do with what he thought of female choice and sexual selection and more to do with his recent debate with Dobzhansky and Huxley over definitions of "fitness." Whereas Mayr and Huxley insisted that fitness connoted a form of physical adaptedness, Dobzhansky and other population geneticists countered that only genetic contribution to the next generation mattered—to Mayr and Huxley, fitness was abso- lute; to Dobzhansky, fitness was relative. See my elaboration of this debate in chapter 5.

78. Adrienne Zihlman, "*Sexual Selection and the Descent of Man, 1871–1971*" (review), *American Anthropologist* 67, no. 2 (1974).

79. Ibid., 476.

80. Peter O'Donald, *Genetic Models of Sexual Selection* (Cambridge: Cambridge Univer- sity Press, 1980), 10–22.

81. Peter O'Donald, "The Theory of Sexual Selection," *Heredity* 17 (1962); Ronald Ayl- mer Fisher, *The Genetical Theory of Natural Selection*, 2nd rev. ed. (New York: Dover Publica- tions, 1958).

82. Russell Lande, "Sexual Selection," *Science, New Series* 209, no. 4453 (1980): 268.

83. Ernst Mayr and William B. Provine, eds., *The Evolutionary Synthesis: Perspectives on the Unification of Biology* (Cambridge, MA: Harvard University Press, 1980), 9 (Thomas Hunt Morgan's dismissal of sexual selection), 332 (Huxley's contribution), 345 (passing reference to the topic); *sexual selection* does not appear in the index.

84. Eliot B. Spiess, "Evolution of Mating Preferences," *Evolution* 34, no. 6 (1980): 1227.

85. Author's interview with Claudine Petit, Dec. 2002; Claudine Petit, "Le déterminisme génétique et psycho-physiologique de la compétition sexuelle chez *Drosophila melanogaster*," *Bulletin Biologique de la France et de la Belgique* 92 (1958).

86. Spiess cited Bateman's 1948 paper, the work of ethologists Margaret Bastock and Aubrey Manning, David Merrell's insistence on female-choice tests for reproductive isolation, and his own research. Eliot B. Spiess, "Do Female Flies Choose Their Mates?" *American Naturalist* 119, no. 5 (1982), 675.

87. Patrick Bateson, ed., *Mate Choice* (Cambridge: Cambridge University Press, 1983).

88. Mark Kirkpatrick, "Revenge of the Ugly Duckling," *Evolution* 38, no. 3 (1984), 704–5.

89. Mayr to Mark Kirkpatrick, July 9, 1984, Ernst Mayr Papers, General Correspondence, 1952–1987, HUGFP 74.7, box 32, folder 1382, HU.

90. Ernst Mayr, *The Growth of Biological Thought: Diversity, Evolution, and Inheritance* (Cambridge, MA: Belknap Press of Harvard University Press, 1982), 596.

91. Author's interview with John Maynard Smith, Dec. 2002.

92. John Alcock, *Animal Behavior: An Evolutionary Approach* (Sunderland, MA: Sinauer Associates, 1975); 2nd ed. (1979), 211; 3rd ed. (1984), 344; 4th ed. (1989), 398; 5th ed. (1993), 402; and 6th ed. (1998), 433.

93. When Helena Cronin published her philosophically oriented *Ant and the Peacock* in 1991, Paul Griffiths remarked in a review that despite the hubbub over her description of the history and current state of affairs in studies of altruism, no one had criticized her history of sexual selection. Cronin's book is an excellent exploration of the philosophical stakes inherent to sexual selection research in the 1970s and 1980s, and she was primarily interested in investigating how past debates in biology could be used to illuminate the present. However, as Griffith notes, such Whig histories only work when everyone agrees on the current state of affairs. What is most remarkable about this episode, then, is that by the early 1990s, all reviewers *did* agree on the current state of affairs on sexual selection. Paul E. Griffiths, "The Cronin Controversy," *British Journal for the Philosophy of Science* 46 (1995): 123, 35.

94. Miller, "How Mate Choice Shaped Human Nature."

95. Ronald Aylmer Fisher, "Positive Eugenics," *Eugenics Review* 9 (1917); Fisher, *Genetical Theory*; Theodosius Dobzhansky, "Experiments on Sexual Isolation in Drosophila. III. Geographic Strains of *Drosophila sturtevanti*," *Proceedings of the National Academy of Sciences of the United States of America* 30, no. 11 (1944); Theodosius Dobzhansky and George Streisinger, "Experiments on Sexual Isolation in Drosophila. II. Geographic Strains of *Drosophila prosaltans*," *Proceedings of the National Academy of Sciences of the United States of America* 30, no. 11 (1944); H. J. Muller, "The Darwinian and Modern Conceptions of Natural Selection," *Proceedings of the American Philosophical Society* 93 (1949); Sewall Wright, "Evolution in Mendelian Populations," *Genetics* 16 (1931).

96. Stephen Jay Gould's last and most extensive exploration of the "hardening" of evolutionary theory was published posthumously. Stephen Jay Gould, *Structure of Evolutionary Thought* (Cambridge, MA: Harvard University Press, 2004), chap. 7, "The Modern Synthesis as a Limited Consensus."

97. John N. Thompson, "Concepts of Coevolution," *Trends in Ecology and Evolution* 4, no. 6 (1989); Michael Ruse and David Sepkoski, eds., *Paleontology at the High Table: The Emergence of Paleobiology as an Evolutionary Discipline* (Chicago: University of Chicago Press, 2007).

98. Arthur Caplan, *The Sociobiology Debate: Readings on Ethical and Scientific Issues* (New York: Harper and Row, 1978); Richard Lewontin, "Sociobiology—A Caricature of Darwinism," *PSA: Proceedings of the Biennial Meeting of the Philosophy Society of America*, no. 2, Symposia and Invited Papers (1976): 22.

99. My analysis builds on Lisa Lloyd's division of the "units of selection" debates into four separate questions: "What is the interactor? What is the replicator? What is the beneficiary? and What entity manifests any adaptations resulting from evolution by selection?" Biologists have used "reductionism" as a criticism directed at opponents who have aimed their analysis of any of these questions at an inappropriately low level of biological organization. Elisabeth A. Lloyd, "Units and Levels of Selection: An Anatomy of the Units of Selection Debate," in *Thinking about Evolution: Historical, Philosophical, and Political Perspectives*, ed. Rama S. Singh, Costas B. Krimbas, Diane B. Paul, and John Beatty (Cambridge: Cambridge University Press, 2001).

Conclusion

1. Vernon Kellogg, *Darwinism To-day: A Discussion of Present-day Scientific Criticism of the Darwinian Selection Theories, together with a Brief Account of the Principal Other Proposed Auxiliary and Alternative Theories of Species-Forming* (New York: Henry Holt and Co., 1907); Vernon Kellogg, *Evolution* (New York, London: D. Appleton and Co., 1924); Thomas Hunt Morgan, *A Critique of the Theory of Evolution* (Princeton, NJ: Princeton University Press, 1916); Thomas Hunt Morgan, *Evolution and Genetics* (Princeton, NJ: Princeton University Press, 1925).

2. David M. Buss, ed., *The Handbook of Evolutionary Psychology* (Hoboken, NJ: John Wiley and Sons, 2005); Lorraine Daston and Fernando Vidal, eds., *Moral Authority of Nature* (Chicago: University of Chicago Press, 2004); Michael T. Ghiselin, "Darwin and Evolutionary Psychology," *Science* 179, no. 4077 (1973).

3. John Beatty, "Ecology and Evolutionary Biology in the War and Post-war Years," *Journal of the History of Biology* 21 (1988); Joseph Allen Cain, "Common Problems and Cooperative Solutions: Organizational Activity in Evolutionary Studies," *Isis* 84 (1993); Joseph Allen Cain, "Ernst Mayr as Community Architect: Launching the Society for the Study of Evolution and the Journal *Evolution*," *Biology and Philosophy* 9 (1994); Joseph Allen Cain and Michael Ruse, eds., *Descended from Darwin: Insights into the History of American Evolutionary Studies, 1900–1970* (Philadelphia: American Philosophical Society, 2009); Jean Gayon, *Darwinism's Struggle for Survival: Heredity and the Hypothesis of Natural Selection* (New York: Cambridge University Press, 1998); Ernst Mayr and William B. Provine, eds., *The Evolutionary Synthesis:*

Perspectives on the Unification of Biology (Cambridge, MA: Harvard University Press, 1980); Vassiliki Betty Smocovitis, "Disciplining Evolutionary Biology: Ernst Mayr and the Founding of the Society for the Study of Evolution and *Evolution* (1939–1950)," *Evolution* 48, no. 1 (1994); Vassiliki Betty Smocovitis, *Unifying Biology: The Evolutionary Synthesis and Evolutionary Biology* (Princeton, NJ: Princeton University Press, 1996); Vassiliki Betty Smocovitis, "Unifying Biology: The Evolutionary Synthesis and Evolutionary Biology," *Journal of the History of Biology* 25 (1992). What is needed, then, is a polyvalent narrative that includes all the seemingly disparate accounts of the synthesis period, however constituted; Joseph Allen Cain, "A Matter of Perspective: Multiple Readings of George Gaylord Simpson's *Tempo and Mode in Evolution*," *Archives of Natural History* 30, no. 1 (2003).

4. William B. Provine, *Origins of Theoretical Population Genetics* (Chicago: University of Chicago Press, 1971; reprint, 2001); William B. Provine, "The Role of Mathematical Population Geneticists in the Evolutionary Synthesis of the 1930s and 40s," *Studies in the History of Biology* 2 (1978).

5. Theodosius Dobzhansky, *Genetics and the Origin of Species* (New York: Columbia University Press, 1937); Mayr and Provine, *Evolutionary Synthesis.*

6. Sol Tax, ed., *Evolution after Darwin,* vol. 1: *The Evolution of Life: Its Origin, History, and Future* (Chicago: University of Chicago Press, 1960); Sol Tax, ed., *Evolution after Darwin,* vol. 2: *The Evolution of Man: Man, Culture, and Society* (Chicago: University of Chicago Press, 1960); Sol Tax and Charles Callender, eds., *Evolution after Darwin,* vol. 3: *Issues in Evolution: The University of Chicago Centennial Discussions* (Chicago: University of Chicago Press, 1960).

7. Vassiliki Betty Smocovitis, "The 1959 Darwin Centennial Celebration in America," *Osiris, 2nd Series* 14, Commemorative Practices in Science: Historical Perspectives on the Politics of Collective Memory (1999).

8. Stephen Jay Gould, *Structure of Evolutionary Thought* (Cambridge, MA: Harvard University Press, 2004), chap. 7, "The Modern Synthesis as a Limited Consensus."

9. Stephen Jay Gould, "The Spandrels of San Marco and the Panglossian Paradigm," *Proceedings of the Royal Society of London, Series B Biological Sciences* 205 (1979); David Sepkoski, "Stephen Jay Gould, Jack Sepkoski, and the 'Quantitative Revolution' in American Paleobiology," *Journal of the History of Biology* 38 (2005); David Sepkoski and Michael Ruse, *The Paleobiological Revolution: Essays on the Growth of Modern Paleontology* (Chicago: University of Chicago Press, 2009).

10. Mayr and Provine, *Evolutionary Synthesis.*

11. Trivers had ignored Mayr's plea for him to keep his mount shut (see chapter 6). Developmental biologists are typically thought to have welcomed an evolutionary focus only in the 1980s with the rise of evo-devo. "Transcript of the Evolutionary Synthesis Conference," Ernst Mayr, Evolutionary Synthesis Papers, B/M45it, folder 13.1, p. 46, Manuscript Collection, APS; Scott F. Gilbert, J. M. Opitz, and Rudolph A. Raff, "Resynthesizing Evolutionary and Developmental Biology," *Developmental Biology* 173, no. 2 (1996); cf. Gregory Davis, Michael Dietrich, and David Jacobs, "Homeotic Mutants and the Assimilation of Developmental Genetics into the Evolutionary Synthesis, 1915–1952," in Cain and Ruse, eds., *Descended from Darwin.*

12. Helena Cronin, *The Ant and the Peacock: Altruism and Sexual Selection from Darwin*

to Today (Cambridge: Cambridge University Press, 1991); Daniel J. Kevles, *In the Name of Eugenics: Genetics and the Uses of Human Heredity* (New York: Alfred A. Knopf, 1985); Provine, *Origins of Theoretical Population Genetics.*

13. Howard Kaye, *The Social Meaning of Modern Biology: From Social Darwinism to Sociobiology* (New Haven, CT: Yale University Press, 1986), 39.

14. Ronald Aylmer Fisher, "Positive Eugenics," *Eugenics Review* 9 (1917); Ronald Aylmer Fisher, "Some Hopes of a Eugenist," *Eugenics Review* 5 (1914); Ronald Aylmer Fisher, "The Evolution of Sex Preference," *Eugenics Review* 7 (1915). The mathematical formulation came in Fisher's *The Genetical Theory of Natural Selection* (Oxford: Clarendon Press, 1930).

15. Wendy Kline, *Building a Better Race: Gender, Sexuality, and Eugenics from the Turn of the Century to the Baby Boom* (Berkeley: University of California Press, 2001); Mark A. Largent, *Breeding Contempt: The History of Coerced Sterilization in the United States* (Piscataway, NJ: Rutgers University Press, 2007); Angelique Richardson, *Love and Eugenics in the Late Nineteenth Century: Rational Reproduction and the New Woman* (New York: Oxford University Press, 2003).

16. A second complicating factor may have been Fisher's familiarity with the genetic research of J. S. Huxley and E. B. Ford on *Gammarus chevreuxi*, in which inheritance was complicated by issues of partial penetrance and polygenic traits. Ronald Aylmer Fisher, *The Genetical Theory of Natural Selection: A Complete Variorum Edition*, ed. J. H. Bennett (New York: Oxford University Press, 1999), 55.

17. "The Evolution of Genetics," lecture by R. A. Fisher, delivered Mar. 1953 to the Manchester Literary and Philosophical Society, Fisher Papers, Series 18, AU-BSL.

18. Fisher, *Genetical Theory Variorum*, 174.

19. Mary Bartley, "Conflicts in Human Progress: Sexual Selection and the Fisherian 'Runaway,'" *British Journal for the History of Science* 27 (1994).

20. Ernst Mayr, "Behavior and Systematics," in *Behavior and Evolution*, ed. Anne Roe and George Gaylord Simpson (New Haven, CT: Yale University Press, 1958), 342.

I do not detail here any newly discovered archival collections or provide keen insights into how to uncover previously invisible materials. One of the fascinating features of this story is that my sources were hidden in plain sight. I used a wide variety of archival materials as well as both primary and secondary published sources. This essay is not comprehensive in either case. Most of the references in the Notes are not listed below.

Primary Sources

ARCHIVAL SOURCES

Over the course of this project, I consulted several archives, following the trail of correspondence back and forth across the Atlantic Ocean, and to Australia. One of the delights of archival research is finding letters and references in unexpected places, and most of my exciting tidbits came from letters received and saved by the individuals whose correspondence I was consulting, rather than from something they themselves had written and saved in their files.

The Manuscript Collection at the American Philosophical Society (APS, in Philadelphia) is a fabulous resource for the history of twentieth-century biology. In particular, the APS has continued to support the collection of papers related to the history of genetics, quite broadly conceived in evolutionary, developmental, and organismal terms. At the APS, I consulted the collected papers of Theodosius Dobzhansky (B D65), Charles Cushman Murphy (B M957), Raymond Pearl (MS Coll. P312), Philip Sheppard (MS Coll. 65), George Gaylord Simpson (MS Coll. 31), and Ernst Mayr's materials related to his conferences on the evolutionary synthesis (B M541) and the official transcripts of the conference proceedings (B M541t). Of special note is a file cabinet off to the side in the APS Reading Room, which contains a cross-listed index of all correspondence in the collected papers housed at the APS. In other words, this index allows a researcher to look up the name of a scientist in whom she is interested and find all of the collected papers with files containing letters from or to that individual. As of December 2008, this file cabinet had not been digitized and remains well worth consulting in person.

Because Ernst Mayr was a prolific correspondent and self-conscious of his authoritative position within twentieth-century evolutionary biology, from early in his career he began to save not only the letters he received from his colleagues, but also carbon copies of many of the letters he sent (Harvard University Archives, Cambridge, Massachusetts). This makes his correspondence a worthwhile stop on any tour of the history of evolutionary biology in the twentieth century. Harvard University Archives' online index to the correspondence contains a list of all letters (author, addressee, and some topical references) in Mayr's General Correspondence, 1931–1952 (HUGFP 14.7), General Correspondence, 1952–1987 (HUGFP 74.7), and Miscellaneous or Anonymous Correspondence, 1920–1993 (HUGFP 74.10). Additionally, I consulted his collected correspondence with Julian Huxley, 1937–1974 (HUGFP 14.15) and Konrad Lorenz and Nikolaas Tinbergen, 1953–1982 (HUGFP 14.17). In Mayr's collected correspondence, I found a trove of letters from Tinbergen to Mayr (many handwritten) that are not included among Tinbergen's own papers at Oxford.

Julian Sorrell Huxley's Papers (MS 50) are located at the Woodson Research Center's Fondren Library of Rice University (in Houston, Texas) and contain his correspondence, personal diaries and notebooks, manuscript drafts, and other material he collected during his travels. Huxley's papers were also useful in learning more about Tinbergen through their correspondence.

The Bodleian Library at the University of Oxford, Department of Western Manuscripts, houses the collected papers and correspondence of former professors, including those of Nikolaas Tinbergen (NCUACS 27.3.91), Edmund Brisco Ford (NCUACS 14.7.89), and Cyril Dean Darlington (CSAC 106.3.85). Ford's papers contain several gems. My favorite was his correspondence with Julian Huxley about their collaborative research on *Gammarus chevreuxi*—a brave but ultimately unsuccessful attempt to establish *Gammarus* as a British model organism to rival American geneticists' monopoly on *Drosophila*.

At the American Museum of Natural History (AMNH, in New York City), Special Collections is housed in the central Research Library and contains materials related to the administration of each department, as well as biographical information of past employees, some personal papers, press releases, and an amazing collection of photographs and films. In addition, each individual department may maintain its own archival records. The archival collection in the Department of Herpetology, for example, contains the collected papers and correspondence of many of its past curators, including Gladwyn Kingsley Noble and Charles M. Bogert.

The collected papers of Ronald Aylmer Fisher (MSS 0013) are housed at the Barr Smith Library, University of Adelaide, Special Collections (in Adelaide, Australia), where he moved just before he died. Much, but not all, of this collection has now been digitized and made available through the R. A. Fisher Digital Archive (http://digital.library.adelaide.edu.au/coll/special/fisher/index.html). Fisher's papers contain a far more complete record of his correspondence with E. B. Ford (dating from 1930 to 1961) than do Ford's own papers.

As I was already in Australia to look at Fisher's correspondence, I also consulted the collected papers of Alan John (Jock) Marshall (MS 7132). Although his saved correspondence is not extensive, his papers contain an excellent record of the many popular articles he wrote, as well as several transcripts of radio programs in which he participated. The Marshall Papers are held at the National Library of Australia (in Canberra).

The official correspondence and program files for the National Research Council, Division of Medical Sciences, Committee for Research in Problems of Sex (NRC-CRPS), 1920–1965, are housed at the National Academies Archives (in Washington, DC). Before the creation of the National Science Foundation, the National Research Council funded much of the research in psychobiology in the United States, including many research projects at the AMNH in the 1930s and 1940s. To learn more about the NRC-CRPS, see Sophie D. Aberle and George W. Corner's *Twenty-five Years of Sex Research: History of the National Research Council Committee for Research in Problems of Sex, 1922–1947* (Philadelphia: W. B. Saunders, 1953) and Adele E. Clarke's *Disciplining Reproduction: Modernity, American Life Sciences, and the "Problems of Sex"* (Berkeley: University of California Press, 1998).

As a final note, digital archival projects are becoming increasingly important resources. Of particular note is the "Complete Work of Charles Darwin Online," directed by John van Wyhe and currently hosted by the Centre for Research in the Arts, Social Sciences and Humanities (CRASSH) at the University of Cambridge (http://darwin-online.org.uk).

ORAL HISTORY SOURCES

First-hand accounts of this exciting period in the history of twentieth-century evolutionary biology helped enrich my understanding of biological practice and the importance of long-term friendships in science, in ways unavailable through the written word. After receiving clearance from the Human Subjects Review Board at the University of Wisconsin–Madison, I interviewed Lee Ehrman, John Maynard Smith, Claudine Petit, Eliot Spiess, and Robert Trivers. I thoroughly enjoyed every minute, and thank them for sharing their memories and stories.

JOURNALS

To familiarize myself with the history of animal behavior, evolution, and especially courtship, I found it extremely helpful to browse through complete runs of biological journals key to the history of evolution and animal behavior, from their first printing to about 1980. As I read, I paid attention both to the networks of people associated with the journal (the identities of the editors, the board of directors, and the published authors) and to the kinds of research published in the journal (which animal subjects, experimental designs, and research communities were represented). For discussions of animal or human courtship in the biological literature, I recommend the *Eugenics Review* (London, 1909–68), *Journal of Genetics* (London, 1910–), *Genetics* (Austin, Texas, 1916–), *Zeitschrift für Tierpsychologie* (Berlin, 1937–85) continued by *Ethology* (1986–), *British Journal of Animal Behaviour* (London, 1953–57) continued as *Animal Behaviour* (London, 1958–), *Evolution* (Lawrence, KS, 1947–), and the early years of *Behavior Genetics* (New York, 1970–).

Secondary Sources

SEXUAL SELECTION

In terms of scholarship by historians on the long durée of sexual selection, of particular note is Helena Cronin's *The Ant and the Peacock: Altruism and Sexual Selection from Darwin to Today* (Cambridge: Cambridge University Press, 1991). Cronin's book is a philosophical investigation of the resonances between debates over sexual selection in the past (Darwin, Wallace, Fisher, Huxley) and debates among biologists in the 1980s. Cronin used her philosophical analysis of these historical debates to shed new light on current disagreements over theories of sexual selection—a self-described "Whig history." Paul Griffiths notes in a meta-review of her book that Whig history works only when scientists agree on the current state of the field ("The Cronin Controversy," *British Journal for the Philosophy of Science* 46 [1995]: 122–38). With this in mind, I believe that the half of Cronin's book devoted to sexual selection succeeded remarkably. Yet, because of her interest in modern biological debates, her analysis could not engage with those aspects of the history that biologists interested in sexual selection had not already identified as crucial to the development of their field (female choice in ethology or population genetics research, for example). With an eye toward the symmetry between Fisher's and Huxley's convictions about human evolution and their theories of animal courtship, historian Mary Bartley completed a valuable dissertation on the history of sexual selection theory and published much of her analysis in two articles: "A Century of Debate: The History of Sexual Selection Theory (1871–1971)" (Ph.D. diss., Cornell University, 1994); "Conflicts in Human Progress: Sexual Selection and the Fisherian 'Runaway,'" *British Journal for the History of Science* 27 (1994): 177–96; and "Courtship and Continued Progress: Julian Huxley's Studies on Bird Behavior," *Journal of the History of Biology* 28 (1995): 91–108. Also useful is an article by Simon J. Frankel (based on his master's thesis at the University of Cambridge): "The Eclipse of Sexual Selection Theory," in *Sexual Knowledge, Sexual Science: The History of Attitudes to Sexuality*, ed. Roy Porter and Mikuláš Teich (Cambridge: Cambridge University Press, 1994), 158–83. A great deal of historical attention has also been directed toward the connection between Darwin's depictions of sex and race in *Descent of Man and Selection in Relation to Sex*. Noteworthy in this regard are Evelleen Richards, "Darwin and the Descent of Woman," in *The Wider Domain of Evolutionary Thought*, ed. David Oldroyd and Ian Langham (London: D. Reidel Publishing Company, 1983), 57–111; Stephen G. Alter, "Race, Language, and Mental Evolution in Darwin's *Descent of Man*," *Journal of the History of the Behavioral Sciences* 43 (2007): 239–55; Jonathan Smith, "Picturing Sexual Selection: Gender and the Evolution of Ornithological Illustration in Charles Darwin's *Descent of Man*," in *Figuring It Out: Visual Languages of Gender in Science*, ed. Bernard Lightman and Ann B. Shteir (Lebanon, NH: University Press of New England, 2007), 85–109; and Adrian Desmond and James Moore, *Darwin's Sacred Cause: How a Hatred of Slavery Shaped Darwin's Views on Human Evolution* (Boston: Houghton Mifflin Harcourt, 2009). No historical analysis of Darwinian sexual selection would be complete without a good hard look at Robert Richard's insightful *Darwin and the Emergence of Evolutionary Theories of Mind and Behavior* (Chicago: University of Chicago Press, 1987).

The book that first sparked my interest in sexual selection, as an undergraduate biology major, was Michael J. Ryan's *Túngara Frog: A Study in Sexual Selection and Communication* (Chicago: University of Chicago Press: 1985). It was out of print by then, and further research by Ryan had already led him to reframe his argument about vocalizations in frogs and sexual selection, but I found his book a gripping narrative nonetheless. Since then, several biologists have written books about sexual selection aimed at a more general audience. Probably the most successful of these are Matt Ridley's *The Red Queen: Sex and the Evolution of Human Nature* (New York: HarperCollins, 2003) and Olivia Judson's *Dr. Tatiana's Sex Advice to All Creation: The Definitive Guide to the Evolutionary Biology of Sex* (New York: Henry Holt & Company, 2002). Recently, sexual selection in animals has received a thorough-going critique at the hands of noted evolutionary biologist Joan Roughgarden. Roughgarden argues that all theories of sexual selection presuppose binary sex (male, female), yet this condition is less common in the animal kingdom than most people suppose; see her *Evolution's Rainbow: Diversity, Gender, and Sexuality in Nature and People* (Berkeley: University of California Press, 2004), and *The Genial Gene: Deconstructing Darwinian Selfishness* (Berkeley: University of California Press, 2009).

GENERAL WORKS ON THE HISTORY OF BIOLOGY AND EVOLUTION IN THE TWENTIETH CENTURY

A variety of books provide useful overviews of the history of evolution in the twentieth century. A great place to start is Garland Allen's classic *Life Sciences in the Twentieth Century* (New York: John Wiley & Sons, 1975). More recent works include the following: Keith Benson, Jane Maienschein, and Ronald Rainger, eds., *The Expansion of American Biology* (New Brunswick, NJ: Rutgers University Press, 1991); Peter Bowler, *Evolution: The History of an Idea* (Berkeley: University of California Press, 2003); Soraya de Chadarevian, *Designs for Life: Molecular Biology after World War II* (Cambridge: Cambridge University Press, 2002); Michel Morange, *A History of Molecular Biology*, trans. Matthew Cobb (Cambridge, MA: Harvard University Press, 1998); Ronald Rainger, Keith Benson, and Jane Maienschein, eds., *The American Development of Biology* (Philadelphia: University of Pennsylvania Press, 1988); Philip J. Pauly, *Biologists and the Promise of American Life: From Merriwether Lewis to Alfred Kinsey* (Princeton, NJ: Princeton University Press, 2000); and Jan Sapp, *Evolution by Association: A History of Symbiosis* (New York: Oxford University Press, 1994). Of particular interest for the history of the evolutionary synthesis are Jean Gayon's *Darwinism's Struggle for Survival: Heredity and the Hypothesis of Natural Selection* (Cambridge: Cambridge University Press, 1998), Vassiliki Betty Smocovitis's *Unifying Biology: The Evolutionary Synthesis and Evolutionary Biology* (Princeton, NJ: Princeton University Press, 1996), and Joe Cain and Michael Ruse's recent volume, *Descended from Darwin: Insights into the History of Evolutionary Studies, 1900–1970*, vol. 99, pt. 1 (Philadelphia: American Philosophical Society, Transactions of the American Philosophical Society, 2009). With the caveat that Ernst Mayr had a stake in the future of evolutionary research, I also recommend the participants' history of the evolutionary synthesis provided in the volume edited by Mayr and William Provine, *The Evolutionary Synthesis: Perspectives on the Unification of Biology* (Cambridge, MA: Harvard University Press, 1980), and Mayr's *The Growth of Biological Thought: Diversity, Evolution, and Inheritance* (Cambridge, MA: Belknap Press of Harvard University Press, 1982). Another important and thorough overview of evolutionary thought, and a participant's history from a different perspective, is Stephen Jay Gould's *Structure of Evolutionary Thought* (Cambridge, MA: Harvard University Press, 2004).

ANIMAL BEHAVIOR IN THEORY AND IN PRACTICE, BOTH LABORATORY AND FIELD

Historians have typically approached the issue of scientific practice in the life sciences from a laboratory perspective. See Adele E. Clarke and Joan H. Fujimura, eds., *The Right Tools for the Job: At Work in Twentieth-Century Life Sciences* (Princeton, NJ: Princeton University

Press, 1992); Angela Creager, *Life of a Virus: Tobacco Mosaic Virus as an Experimental Model* (Chicago: University of Chicago Press, 2002); Robert Kohler, *Lords of the Fly:* Drosophila *Genetics and the Experimental Life* (Chicago: University of Chicago Press, 1994); and Karen A. Rader, *Making Mice: Standardizing Animals for American Biomedical Research, 1900–1955* (Princeton, NJ: Princeton University Press, 2004).

When historians of science have addressed animal-human interactions in a field context, they have mostly done so from a philosophical or cultural perspective. In these historical analyses, animals act as methodological windows through which to see the inner workings of human cultures or as mirrors that reflect the historical relationships of labor, race, or gender in a society. These include Molly H. Mullin, "Mirrors and Windows: Sociocultural Studies of Human-Animal Relationships," *Annual Review of Anthropology* 28 (1999): 201–24; and Lorraine Daston and Gregg Mitman, eds., *Thinking with Animals: New Perspectives on Anthropomorphism* (New York: Columbia University Press, 2005). Keen analysis along these lines reveals much about culturally specific notions of imperial power, class, religion, nature, and morality. See Donna Haraway, *Primate Visions: Gender, Race, and Nature in the World of Modern Science* (New York: Routledge, 1990) and *The Companion Species Manifesto: Dogs, People, and Significant Otherness* (Cambridge, UK: Prickly Pear, 2003); Nicholas Jardine, James Secord, and Emma Spary, eds., *Cultures of Natural History* (Cambridge: Cambridge University Press, 1996); Harriet Ritvo, *The Animal Estate: The English and Other Creatures in the Victorian Age* (Cambridge, MA: Harvard University Press, 1987); Peter Singer, *The Expanding Circle: Ethics and Sociobiology* (New York: Farrar, Straus & Giroux, 1981); Frans de Waal, Robert Wright, Christine M. Korsgaard, Philip Kitcher, and Peter Singer, *Primates and Philosophers: How Morality Evolved* (Princeton, NJ: Princeton University Press, 2006); and Brett Walker, *The Lost Wolves of Japan* (Seattle: University of Washington Press, 2005).

Recently, historians of science have increasingly turned their attention to the theoretical and practical approaches of twentieth-century animal behavior research outside the laboratory context, in the "field" broadly construed. These books combine an appreciation of animals as research objects in twentieth-century science with a close analysis of the evolutionary, linguistic, and cultural relationships between animals and humans. Examples are Richard W. Burkhardt, Jr., *Patterns of Behavior: Konrad Lorenz, Niko Tinbergen, and the Founding of Ethology* (Chicago: University of Chicago Press, 2005); Eileen Crist, *Images of Animals: Anthropomorphism and Animal Mind* (Philadelphia: Temple University Press, 2000); Rebecca Lemov, *World as Laboratory: Experiments with Mice, Mazes, and Men* (New York: Hill and Wang, 2005); Gregory Radick, *The Simian Tongue: The Long Debate about Animal Language* (Chicago: University of Chicago Press, 2007); Edmund Russell, *War and Nature: Fighting Humans and Insects with Chemicals from World War I to Silent Spring* (Cambridge: Cambridge University Press, 2001); and Charlotte Sleigh, *Six Legs Better: A Cultural History of Myrmecology* (Baltimore: Johns Hopkins University Press, 2007).

Robert Kohler is one of the few historians of science to analyze the traffic of methods, ideas, and ideals between the laboratory and field cultures of ecologists and evolutionary biologists; see his *Landscapes and Labscapes: Exploring the Lab-Field Border in Biology* (Chicago: University of Chicago Press, 2002). What clearly emerges from his narrative is that, as places of research, the laboratory and the field provide complementary forms of authenticity. Animal behavior is a fascinating discipline within the life sciences in part because practitioners have long recognized and embraced both forms of research as central to their enterprise.

WOMEN, GENDER, SCIENCE

Feminist analyses have enriched our understanding of the cultures and processes of the sciences and expanded our definitions of what kinds of activities are central to the scientific enterprise, and in doing so have enlarged our roster of who counts as doing scientific research. See Anne Fausto-Sterling, *Myths of Gender: Biological Theories about Men and Women* (New York: Basic Books, 1985), and *Sexing the Body: Gender Politics and the Construction of Sexuality* (New York: Basic Books, 2000); Barbara T. Gates, *Kindred Nature: Victorian and Edwardian Women Embrace the Living World* (Chicago: University of Chicago Press, 1998); Barbara T. Gates and Ann B. Shteir, eds., *Natural Eloquence: Women Reinscribe Science* (Madison: University of Wisconsin Press, 1997); Evelyn Fox Keller, *A Feeling for the Organism: The Life and Work of Barbara McClintock* (San Francisco: W. H. Freeman and Co., 1983); Sally Gregory Kohlstedt, Barbara Laslett, Helen Longino, and Evelyn Hammonds, eds., *Gender and Scientific Authority* (Chicago: University of Chicago Press, 1996); Sally Gregory Kohlstedt, "Nature Not Books: Scientists and the Origins of the Nature Study Movement in the 1890s," *Isis* 96, no. 3 (2005): 324–52; Sally Gregory Kohlstedt and Mark R. Jorgensen, "'The Irrepressible Woman Question': Women's Response to Darwinian Evolutionary Ideology," in *Disseminating Darwinism: The Role of Place, Race, Religion, and Gender*, ed. Ronald L. Numbers and John Stenhouse (Cambridge: Cambridge University Press, 1999); Elisabeth Lloyd, *The Case of the Female Orgasm: Bias in the Science of Evolution* (Cambridge, MA: Harvard University Press, 2005); Margaret Rossiter, *Women Scientists in America: Struggles and Strategies to 1940* (Baltimore: Johns Hopkins University Press, 1982); Cynthia Eagle Russett, *Sexual Science: The Victorian Construction of Womanhood* (Cambridge, MA: Harvard University Press, 1989); and Londa Schiebinger, *The Mind Has No Sex? Women in the Origins of Modern Science* (Cambridge, MA: Harvard University Press, 1989).

EUGENICS AND GENETICS

The story of eugenic policies and research on genetics in humans has continued to garner the attention of historians. A good place to start in this literature is with Daniel Kevles, *In the Name of Eugenics: Genetics and the Uses of Human Heredity* (New York: Alfred A. Knopf, 1985); and Diane Paul, *Controlling Human Heredity: 1865 to the Present* (Amherst, MA: Humanities Books, 1998). More recent and equally engaging work on the history of eugenics includes the following: John Carson, *The Measure of Merit: Talents, Intelligence, and Inequality in the French and American Republics, 1750–1940* (Princeton, NJ: Princeton University Press, 2007); Wendy Kline, *Building a Better Race: Gender, Sexuality, and Eugenics from the Turn of the Century to the Baby Boom* (Berkeley: University of California Press, 2001); Mark Largent, *Breeding Contempt: The History of Coerced Sterilization in the United States* (New Brunswick, NJ: Rutgers University Press, 2007); Angelique Richardson, *Love and Eugenics in the Late Nineteenth Century: Rational Reproduction and the New Woman* (New York: Oxford University Press, 2003); and Leila Zenderland, *Measuring Minds: Henry Herbert Goddard and the Origins of American Intelligence Testing* (Cambridge: Cambridge University Press, 2001).

THE AUTHORITY OF NATURE

In science as in daily life, claims that something is "natural" carry great weight. Such claims are culturally specific and extremely powerful in dictating rules of normality. The following books were helpful as I thought through these issues with regard to the history of female choice and sexual selection: William Cronon, *Uncommon Ground: Rethinking the Human Place in Nature* (New York: W. W. Norton, 1995); Lorraine Daston and Gregg Mitman, eds., *Thinking with Animals: New Perspectives on Anthropomorphism* (New York: Columbia University Press, 2005); Lorraine Daston and Fernando Vidal, eds., *The Moral Authority of Nature* (Chicago: University of Chicago Press, 2004); Donna Haraway, *Primate Visions: Gender, Race, and Nature in the World of Modern Science* (New York: Routledge, 1989); Gregg Mitman, *The State of Nature: Ecology, Community, and American Social Thought, 1900–1950* (Chicago: University of Chicago Press, 1992); and Miriam G. Reumann, *American Sexual Character: Sex, Gender, and National Identity in the Kinsey Reports* (Berkeley: University of California Press, 2005).

Page numbers in *italics* indicate illustrations.